A Dynamic Model of Multilingualism

MULTILINGUAL MATTERS SERIES
Series Editor: Professor John Edwards, *St Francis Xavier University, Antigonish, Nova Scotia, Canada*

Other Books in the Series
Beyond Bilingualism: Multilingualism and Multilingual Education
   Jasone Cenoz and Fred Genesee (eds)
Can Threatened Languages be Saved?
   Joshua Fishman (ed.)
Community and Communication
   Sue Wright
Identity, Insecurity and Image: France and Language
   Dennis Ager
Language and Society in a Changing Italy
   Arturo Tosi
Language Attitudes in Sub-Saharan Africa
   Efurosibina Adegbija
Language, Ethnicity and Education
   Peter Broeder and Guus Extra
Language Planning in Malawi, Mozambique and the Philippines
   Robert B. Kaplan and Richard B. Baldauf, Jr. (eds)
Language Planning in Nepal, Taiwan and Sweden
   Richard B. Baldauf, Jr. and Robert B. Kaplan (eds)
Language Planning: From Practice to Theory
   Robert B. Kaplan and Richard B. Baldauf, Jr. (eds)
Language Reclamation
   Hubisi Nwenmely
Linguistic Minorities in Central and Eastern Europe
   Christina Bratt Paulston and Donald Peckham (eds)
Motivation in Language Planning and Language Policy
   Dennis Ager
Multilingualism in Spain
   M. Teresa Turell (ed.)
Quebec's Aboriginal Languages
   Jacques Maurais (ed.)
The Other Languages of Europe
   Guus Extra and Durk Gorter (eds)
The Step-Tongue: Children's English in Singapore
   Anthea Fraser Gupta
A Three Generations – Two Languages – One Family
   Li Wei

Please contact us for the latest book information:
Multilingual Matters, Frankfurt Lodge, Clevedon Hall,
Victoria Road, Clevedon, BS21 7HH, England
http://www.multilingual-matters.com

MULTILINGUAL MATTERS 121
Series Editor: John Edwards

A Dynamic Model of Multilingualism
Perspectives of Change in Psycholinguistics

Philip Herdina and Ulrike Jessner

MULTILINGUAL MATTERS LTD
Clevedon • Buffalo • Toronto • Sydney

Library of Congress Cataloging in Publication Data
Herdina, Philip
A Dynamic Model of Multilingualism: Perspectives of Change in Psycholinguistics/
Philip Herdina and Ulrike Jessner.
Multilingual Matters: 121
Includes bibliographical references and index.
1. Multilingualism–Psychological aspects. 2. Psycholinguistics. 3. Second language acquisition. 4. Systems theory. I. Jessner, Ulrike. II. Title. III. Multilingual Matters (Series): 121.
P115.4.H47 2002
306.44'6–dc21   2001026622

British Library Cataloguing in Publication Data
A catalogue entry for this book is available from the British Library.

ISBN 1-85359-468-7 (hbk)
ISBN 1-85359-467-9 (pbk)

Multilingual Matters Ltd
UK: Frankfurt Lodge, Clevedon Hall, Victoria Road, Clevedon BS21 7HH.
USA: UTP, 2250 Military Road, Tonawanda, NY 14150, USA.
Canada: UTP, 5201 Dufferin Street, North York, Ontario M3H 5T8, Canada.
Australia: Footprint Books, Unit 4/92a Mona Vale Road, Mona Vale, NSW 2103, Australia.

Copyright © 2002 Philip Herdina and Ulrike Jessner.

All rights reserved. No part of this work may be reproduced in any form or by any means without permission in writing from the publisher.

Printed and bound in Great Britain by the Cromwell Press Ltd.

Contents

Preface vii
List of Figures ix
List of Acronyms x

1 Introductory Remarks 1
    Plan of the Book 4
    Overview of Chapters 4

2 Stages in Research on Multilingualism 6
    Double Monolingualism Hypothesis 6
    Earlier Models of Bilingual Representation 7
    Transfer as an Explanation of Linguistic Deficiency 9
    The Impact of the Peal & Lambert Study 14

3 Transfer Reconsidered 19
    Transfer in Research on Bilingualism 20
    Transfer in the Learner System 24
    The Paradox of Transfer 26
    Transfer Phenomena in Multilingual Systems 28

4 Universal Grammar Reviewed 30
    Theory of Competence 30
    Tenets of UG Language Acquisition Theory 32
    UG on SLA 41
    UG on Interim Language and Multilingualism 47

5 Multilingual Proficiency Reassessed 52
    Defining Multilingualism 52
    Current Issues in Multilingualism Research 58
    Multilingual Variation 69
    Arguments for a Dynamic View 74

6 A Dynamic Model of Multilingualism Developed 76
    Introduction to Dynamic Systems Approaches 76
    Building DMM 85

	Key Factors of DMM: Gradual Language Loss and
	Language Maintenance................................ 93
	Hypothetical Assumptions of DMM................ 109
7	A Dynamic Model of Multilingualism Analysed 111
	Conceptual Outline of DMM........................ 111
	Types of Multilingualism According to DMM.......... 117
	How Factors Relate in DMM........................ 125
	Mastering Complexity in DMM..................... 140
8	Holism Defended: A Systems Interpretation............ 144
	Double Monolingualism and Modularity............. 145
	Wholism: The Bilingual View and Multicompetence 148
	Holism and Systems Theory........................ 150
9	Limitations, Conclusions and Outlook 153
	Theoretical Limitations of DMM.................... 153
	Related Concepts of DMM......................... 155
	Questions Raised by DMM......................... 158
	Applications of DMM............................. 159

References 162
Index 180

Preface

We consider this publication to be unique in so far as it tries to combine two fields of research which at first sight might appear to have very little to do with each other. On the one hand there is the theory of dynamic systems which has exerted considerable influence in various scientific fields such as physics, biology and meteorology and on the other hand there is the psycholinguistically and sociolinguistically fascinating field of multilingualism research to which an increasing amount of attention is being paid.

This approach reflects our backgrounds and interests, which – as different as they may be – have contributed to our understanding of the complexities of the issues involved in acquiring and maintaining a multilingual system. Whilst one of the authors stems from a bilingual family background combined with a sustained interest in philosophical and methodological issues, the other has both acquired two further languages and made third language acquisition and trilingualism a focus of her research interests. So while one of the authors wrote some of the first drafts of the text – mainly those dealing with UG criticism and the development of the model – the other concentrated her writing on multilingualism research and put his ideas into relation to the relevant literature in the field. In the final stages of preparation of the manuscript she also transformed and extended the text into a reader-friendly format.

We would like to express our deepest gratitude and appreciation to Jasone Cenoz, David Singleton, and Britta Hufeisen (supported by Martha Gibson and Nicole Marx) who were prepared to read the whole manuscript in its various stages of preparation and provided us with corrective feedback and valuable suggestions. Furthermore we would like to thank various scholars such as Andrew Cohen, Sascha Felix, Manfred Kienpointner, Terence Odlin, Peter Scherfer, to name but a few, and last but not least Jim Cummins for their helpful and critical remarks on some of the ideas presented in the book. Their contributions clearly helped us to bring this book into the form it has found here, whilst the responsibility for errors is obviously entirely our own.

We would also like to thank Helmut Essenschläger without whose expertise as a computer specialist and mathematician this publication would not look the way it does. He designed the graphs and also provided us with valuable comments.

We also wish to express our sincere gratitude to the team at Multilingual Matters for their highly efficient work accompanied by a friendly and helpful style, and in particular to Marjukka Grover for her encouragement and patience. Furthermore, Charlotte Hoffmann who acted as reviewer for Multilingual Matters has to be thanked for her extremely useful comments on the manuscript. And we have to be grateful to the Series Editor, John Edwards, for including it the series. Finally heartfelt thanks go to all those who believed in the positive outcome of this dynamic enterprise!

We can only hope that the reader will sense the fascination inherent in some of the complex problems related to creating a dynamic model of multilingualism and will at least conclude that multilingualism is a subject worthy of further study.

List of Figures

Figure 1	Language range graph.
Figure 2	Ladder model according to UG.
Figure 3	Language acquisition according to the principles and parameters approach.
Figure 4	Resetting model according to UG.
Figure 5	Critical age hypothesis.
Figure 6	Approximative systems model.
Figure 7	Intermediate range model.
Figure 8	Addition versus multiplication.
Figure 9	Changes in the movement of gas particles in a chamber as a function of heat.
Figure 10	Belousov-Zhabotinsky reaction.
Figure 11	Non-linear model of making a non-linear feedback model.
Figure 12	Linear process.
Figure 13	Biological growth.
Figure 14	Gradual language loss.
Figure 15	Scatter graph of language attrition.
Figure 16	Abstract learning curve without any limiting factors.
Figure 17	Actual learning curve influenced by limiting factors.
Figure 18	Growth of LME.
Figure 19	Threshold hypothesis.
Figure 20	Ideal learning curve related to LME.
Figure 21	Approximative homeostatic systems.
Figure 22	The effect of fossilisation on language systems development.
Figure 23	Language proficiency development relative to the learning process with and without monitoring function.
Figure 24	The catalytic effect of MLA on the development of LS_3 or TLA.
Figure 25	Ambilingual balanced bilingualism.
Figure 26	Non-ambilingual balanced bilingualism.
Figure 27	Transitional bilingualism.
Figure 28	Stable dominant bilingualism.
Figure 29a	Learner multilingualism: acquisition phase.
Figure 29b	Learner multilingualism: overall development.
Figure 30	Cumulative measure of multilingual proficiency.
Figure 31	GLE as a function of LAE and LME.
Figure 32	Transition period.
Figure 33	Some individual factors involved in the development of a multilingual system.

List of Acronyms

The following acronyms are used throughout this book.

ACT	Adaptive control of thought.
BICS	Basic interpersonal communication skills.
CAH	Contrastive analysis hypothesis.
CALP	Cognitive/academic language proficiency.
CLI	Crosslinguistic influence.
CLIN	Crosslinguistic interaction.
CS	Codeswitching.
CUP	Common underlying proficiency.
DMM	Dynamic model of multilingualism.
EMM	Enhanced multilingual monitor.
FLA	First language acquisition.
GLE	General language effort.
LAE	Language acquisition effort.
LME	Language maintenance effort.
LS	Language system.
LS_1	First language system.
LS_2	Second language system.
LS_3	Third language system.
LS_P	Primary language system.
LS_S	Secondary language system.
M-factor	Multilingualism factor.
MLA	(Multi)language aptitude/metalinguistic abilities.
MR	Multilingualism research.
PC	Perceived language competence.
PCN	Perceived communicative needs.
PLA	Primary language acquisition.
SLA	Second language acquisition.
SUP	Separate underlying proficiencies.
TLA	Third language acquisition.
UG	Universal Grammar.

Chapter 1
Introductory Remarks

Research interest in the linguistic phenomenon of multilingualism has been increasing over the last decades (see e.g. Edwards, 1994) and it has turned out that multilingualism is not only far more common than originally expected – and therefore of greater sociolinguistic importance – but research into multilingualism is expected to exert considerable influence on linguistic theory. Today an increasing number of opinions are voiced according to which linguistic research should no longer be modelled on the monolingual speaker but should take the bilingual as its point of departure (see Romaine, 1989; Cook, 1993a).

As the majority of the world's population is multilingual, research on linguistics should be centred on the multilingual speaker as the norm, not on the monolingual individual. As Cook (1993b: 245) suggests, basing psycholinguistics on the ideal monolingual speaker (homo monolinguis) in fact constitutes a misunderstanding of at least the majority of native speakers. It has become obvious that our conception of the speaker's language system has to be flexible enough to accommodate the command of more than one language. This requirement necessitates a reevaluation of the dominant conceptions of second language acquisition (henceforth SLA) within a multilingual context (see Kachru, 1994) and the reassessment of the relevant theories on the basis of crosscultural evidence (see Sridhar, 1994). Multilingualism therefore must not only be accepted as the linguistic norm, it must also be realised that it is closely linked to the concepts of personal identity, ethnicity and multiculturalism. We will, however, not touch upon this issue as we do not consider it important for our discussion here (see, e.g. Hamers & Blanc, 1989: 60–83).

The dynamic view outlined in this book attempts to provide a new model and a new set of concepts for the interpretation of psycholinguistic phenomena observed in speakers of more than one language. Although there is no immediate precedent to be found in psycholinguistic (or multilingualism) research – as the psychodynamics of multilingualism have only in part been addressed so far – a systems-theoretic approach as can be found in chaos and complexity theory has been with us in other sciences, such as biology and physics, where it has risen against the trend of reductionism for quite a while now (see Gleick, 1987).

It is probably true to say that recent research into multilingualism has raised rather more questions concerning the acquisition of languages and the nature of human language ability than it has provided answers. Traditional conceptions and explanations of language learning and resulting crosslinguistic effects have thus been called into question. On the basis of these findings it is suggested that our interpretations of language learning, what it is to know a language, etc. urgently require revision. The authors believe that a large number of theories currently discussed in research on language acquisition and multilingualism lack an overall theoretical foundation.

What distinguishes the systems-theoretic view from other approaches to multilingualism? The first claim made by the authors is that the dynamic systems approach, as outlined here, is a novel approach to the field, although the concept of language as a system is by no means a new one (see, e.g. Schweizer, 1979). The novel aspects introduced by the dynamic model include a psycholinguistic focus on the systems-theoretic approach that is based on research on the behaviour of living systems and a dynamic interpretation of the systems model. Whilst other models are satisfied to create a systems interpretation of the field of research, this model makes a point of a dynamic representation of multilingualism. In the dynamic model of multilingualism (henceforth DMM) it does not suffice to determine the relations between various factors within the system, but predictions concerning the typical development of these variables are also attempted.

A systems-theoretic interpretation as suggested by DMM allows a realistic view of the phenomenon of multilingualism, which transcends traditional approaches to multilingualism. This model not only takes into account the methodological insight into the fact that a bilingual speaker is more than two monolingual individuals joined together (see Grosjean, 1985) but also connects SLA research to an originally purely sociolinguistic approach to bi-/multilingualism by integrating knowledge of language learning from various strands of research. Thus on the basis of a closer investigation of multilingualism we obtain a number of research goals that have so far been ignored in linguistic research dominated by the monolingual paradigm. DMM is a psycholinguistic model which sees language change on an individual level as a function of time, that is, a focus is placed on the variability and dynamics of the individual speaker system, an aspect of multilingual learning hitherto largely ignored.

DMM takes an innovative approach to the subject matter in at least two ways: firstly, and in accordance with systems-theoretic principles, DMM views a multilingual speaker as a complex psycholinguistic system comprising individual language systems (LS_1, LS_2, LS_3, etc.) and consequently applies insights gained from the observation of the biological development and behaviour of living organisms to research on multilingualism. Secondly, DMM tries to create an explicit model of multilingualism specifying dependent and independent variables and making predictions about the development of multilingual systems.

Obviously this is a rather ambitious aim at this stage of research. The modelling provided does, however, fulfil the Chomskyan criterion of explicitness (see Chomsky, 1965: 4), a precondition of effective theory development, and also provides a framework within which hitherto inscrutable or insurmountable problems of multilingualism can be addressed. Whilst one might argue that this step is methodologically premature, it is certainly not methodologically naive. It is rather the counterargument that more empirical research has to be conducted prior to the construction of such models that stems from a certain naivety concerning the independence of experience or empirical findings of theoretical preconceptions. As all experience is necessarily theory-laden it is methodologically advisable to create an explicit model of the theoretical preconditions of research conducted in multilingualism, and it is this explicitness that in the authors' view research on multilingualism very often lacks (see Herdina, 1990).

The focus of the book is theoretical and the suggested model is intended to provide an essential and useful framework for future research, both in general and applied linguistics. This book should therefore not be seen as an introduction like, for instance, Baker's most valuable contribution to the field (1996), but as an attempt to introduce a novel perspective serving as a bridge between SLA and multilingualism research, which might be compared to the ecological approach to bilingualism initiated by Haugen (1972).

As our theory claims to cover the topic of multilingualism, the reader will of course be tempted to ask what the authors mean by the term multilingualism. S/he will be acquainted with the accepted distinctions between incipient and fully developed bi-/multilingualism, balanced or unbalanced bilingualism, learner systems and fully developed bilingualism to name a few (see Hoffmann, 1991: 14–18; Lüdi, 1996). Whilst we do not expect the reader to fall into the trap of only accepting the ambilingual speaker as legitimate object of bi-/multilingualism research, there always remains the question of where learner systems or language learning ends and where multilingualism starts. As we hope to show in the course of the book, one of the great advantages of DMM is the fact that we are primarily interested in a holistic conception and understanding of multilingualism, where it is at least initially of little consequence whether the respective language system is available in its incipient form or as a mature one.

Furthermore we assume that the phenomenon of bilingualism on which most of the research available has been conducted is essentially a variant of multilingualism (see Haarmann, 1980: 13) and that many of the findings – but certainly not all – can therefore be generalised to cover all variants of multilingualism. Recent research on third language acquisition (henceforth TLA) (see, e.g. Hufeisen & Lindemann, 1998) has shown that investigations into three languages in contact is a field in its own right and it is one of the aims of the book to model multilingual learning by going beyond the discussion of the contact between two language systems. So, whilst we would like to show how bilingual systems can form the basis of modelling multilingualism, our discussion in this book will also focus on specific features of multilingualism which make it differ from SLA and bilingualism.

Plan of the Book

As the issues raised in this volume are necessarily complex it will prove to be of advantage to provide a brief outline of the topics covered in the book. It will be noted that, methodologically, we adopt a twofold approach. Whilst the Chapters 1 to 5 try to give an overview of the existing problems and theories by pointing out important steps in the history of research on SLA and bilingualism, providing the reader with a grasp of current multilingual issues, the main part of the book expounds both a new concept of multilingualism and the overall theoretical context in which this new interpretation is to be seen. Thus, whilst in the early chapters we necessarily use an inductive approach drawing on numerous and varied sources to back up the arguments, the final chapters are of a more deductive nature, deriving the new model from initial assumptions and hypotheses. The model is unfolded step by step revealing increasing complexities as the details are added. The various stages are illustrated by numerous graphs which are intended to help the reader understand the concepts introduced.

Overview of Chapters

Following some introductory remarks in Chapter 1, Chapter 2 attempts to provide an overview of the main topics raised in multilingualism research, that is a survey of the main issues raised in bilingualism and SLA research. Chapter 3 takes a closer look at the problem of transfer as one of the core questions of multilingualism research, where transfer phenomena are considered in the widest possible terms, and develops some working hypotheses on the basis of the conclusions drawn. Chapter 4 turns to the theories developed by Universal Grammar which have dominated research on multilingualism and language acquisition in order to outline shortcomings and suggest some modifications to the Universal Grammar approach. Chapter 5 takes a closer look at the salient features of multilingualism and interprets the recent findings with a special focus on studies undertaken in TLA research as indicative of the need for a new model of multilingualism. The chapter starts with the discussion of an important problem of current research, that is the clarification of what is to be understood by the command of a language or more than one language. The following two Chapters 6 and 7 represent the core of the book, in which DMM is developed and analysed. Chapter 6 introduces dynamic systems approaches and links them to the development of DMM. In Chapter 7 individual factors of the model are investigated and the greater explanatory power of the dynamic model of multilingualism is demonstrated. Chapter 8 finally looks more closely at the question of holism. It should become clear that DMM presents a fundamentally holistic approach and it has to be pointed out that it differs from w(!)holistic suggestions made in research so far. The closing Chapter 9 focuses on some theoretical issues of the model and presents future research perspectives identified by DMM as well as suggestions for applications in multilingual education.

Chapter 2
Stages in Research on Multilingualism

The question whether the bilingual speaker differs from the monolingual has been at the centre of interest in bilingualism research for almost a century (see Kelly, 1969). Over the last thirty years the number of psycholinguistic studies focusing on the effects of bilingualism, on the intelligence of the child, and on how a bilingual mentally organises the two languages has increased greatly (see Reynolds, 1991b) and with this the linguistic interpretation of SLA and bilingualism have changed considerably. This change has also affected the interpretation of the psychological and educational implications of each interpretation (see Bialystok, 1991a).

We do not intend to give a detailed review of bilingualism research here – because several other authors have done so in detail (see, e.g. Grosjean, 1982; Hakuta, 1986; Appel & Muysken, 1987; Romaine, 1989; Hamers & Blanc, 1989; Hoffmann, 1991; Baker, 1996; Baker & Prys Jones, 1998) – but to provide a description of the research carried out during the last few decades which has developed in the fields we consider relevant for our discussion of psycholinguistic issues. Before we turn to questions of language processing in bilinguals, the role of transfer explanations of learner systems and the impact of the Peal & Lambert study on current research on bilingualism we will first of all focus on the dominant concept of a bilingual as a double monolingual in one person.

Double Monolingualism Hypothesis

Although an increasing number of investigations of language choice, codeswitching and codemixing have indicated that it is common in bilinguals either to use their two languages alternately or to mix them (see, Milroy & Muysken, 1995), the monolingual (or fractional) view – a term introduced by Grosjean (1985) – is still the prevailing concept in studies of bilingualism and SLA. Most research on bilingualism has been based on the view of the bilingual as the sum of two monolinguals in one person with two separate language competences or, in other words, bilinguals being regarded as two monolinguals in one person. Consequently bilingual proficiency has generally been measured against monolingual proficiency.

Early research on bilingualism, conducted by Saer (1922, 1923) and later

Weisgerber (1966), for example, showed the bilingual speaker to be deficiently monolingual or semilingual, even when bilingual competence was acquired prior to the critical period, which in theory should have ensured native speaker competence. We must concede that these findings may partly be attributed to ideological bias. It was initially held that bilingualism would have a detrimental effect not only on language development as such but also on cognitive development in general. In a most comprehensive review Hakuta (1986: 14–44) describes three stages of the history of research on bilingualism and intelligence which he terms the periods of detrimental, neutral and additive effects, thus indicating a developmental pattern in research towards a positive view of bilingualism.

More recent findings, however, still tend to confirm, if not the purported detrimental cognitive effects, then at least the linguistic disadvantage of bilingual children – when compared with monolinguals of the same age – which is generally attributed to the dominance in one of their languages. Countless studies in SLA research seem to prove the linguistic inferiority of bilinguals in comparison with monolinguals. Not only do bilinguals' linguistic resources generally appear to be inferior to those of their monolingual counterparts, there also seems to be ample evidence of interaction between the two language systems, thus making the double monolingualism hypothesis less plausible. It can therefore be noted that as long as bilinguals are measured according to monolingual criteria, they appear to be greatly disadvantaged both in linguistic and cognitive terms (see Baker, 1996: 119).

As bilingualism is a fairly recent research goal in the history of linguistics, it is not surprising that many researchers still tend to draw upon monolingual psycholinguistic theories. Due to the dominance of a mentalist theory of language acquisition, as originally specified by Chomsky (1965), bilingualism is generally interpreted as a kind of double monolingualism, a view of bilingualism that would allow no connection between bilingualism and SLA. In contrast to this we will describe and analyse earlier research on bilingualism and SLA to show up common or complementary aspects in these two related fields of research.

Earlier Models of Bilingual Representation

In the first studies on bilingualism, often carried out in form of diary studies by researchers reporting on the linguistic development of their own children, reflections upon the characteristics of the organisation of a bilingual system and its development were already of special interest. The most well-known and therefore frequently quoted examples are Ronjat (1913), who describes the linguistic development of his son in a German-French family and Leopold (1939–1949), who provided the most detailed report of his daughter Hildegard's simultaneous acquisition of English and German.

The model of language representation in bilinguals that has influenced subsequent research most and therefore turns up in all reviews is Weinreich's (1953) and subsequently Ervin and Osgood's (1954) distinction between compound and coordinate bilingualism in a first attempt to classify bilingual phenomena. These researchers regard a person as a coordinate bilingual, if s/he learns her/his two languages in separate cultural environments, which implies that the vocabularies of the two languages are kept separate with each word having its own specific meaning. A compound bilingual, in contrast, is defined as a person who has learned both languages in the same context. In the first case the development of two different conceptual systems, in the second a fused representation in the brain is assumed. Weinreich adds to this twofold distinction the category of the sub-coordinate bilingual who has a primary set of meanings established through the first language and therefore interprets words of the weaker language through words of the dominant language. Interaction is considered more likely in compound bilinguals than in coordinate bilinguals. This classic model of language encoding in the bilingual brain has stimulated a great deal of discussion on these issues (e.g. Diller, 1970; Perecman, 1984; Wölck, 1987/88).

> The compound-coordinate model of bilingualism was the first statement of the shared and separate store hypothesis of bilingual memory which would provide the basis for most research in bilingual memory over the next 30 years. However, research on the compound-coordinate model had generally been abandoned by the end of the 1960s. Perhaps this was due to problems within both model and research, but it was probably also caused by the general excitement within psychology about models based on information processing frameworks. These models linked questions about memory organisation to those of the nature of representation. (Keatley, 1992: 17)

Hamers and Blanc (1989: 8), for instance, argue that for them the distinction has to do with a difference of cognitive organisation and not with a difference in the degree of competence, or a difference in the age or context of acquisition although there is a high correlation between the type of cognitive organisation, age and context of acquisition.

Since the introduction of the compound/coordinate distinction questions concerning the nature of the development of a bilingual's language systems and their possible interaction have been at the centre of interest. Are bilingual systems merged or separate? Do they interact with each other and if so how? The question of mental representation, the one-versus-two-systems hypothesis or the independence-versus-interdependence hypothesis is still one of the key issues in psycholinguistic research (see Bialystok & Hakuta, 1994: 110–119; Baker, 1996: 125–127).

Like many researchers afterwards, Leopold and Ronjat, for instance, expressed their opinions on this issue and argued that the child has only one lexical system at an early age. Evidence suggests that simultaneous acquisition of bilingualism in childhood is linked with fused systems in the earliest stages. Later distinct linguistic systems for the two languages involved will develop (see Romaine, 1989: 81). There is still a certain controversy concerning the differentiation of language systems in a bilingual child but also concerning lateralisation (see Vaid & Hall, 1991) and most recently research on the cortical representation of native and second languages by Kim *et al.* (1997) has even shown that second languages acquired in adulthood are spatially separated from native languages.

Transfer as an Explanation of Linguistic Deficiency

The obvious problem research on SLA and bilingualism has had to deal with is that of underachievement or partial achievement. Having rejected the early view that bilinguals grow up to become cognitively deficient, research still faces the problem that L2 learners have been shown to be deficient, for instance, in syntactic comprehension, text information recall and word comprehension (see Cook, 1993a: 110). SLA research has clearly proved this type of multilingual to be disadvantaged or underachieving, when compared with monolingual speakers. This phenomenon obviously requires explanation, if we assume the multilingual speaker to be equally intelligent or gifted as the monolingual speaker.

Various explanations have been offered as to why second language learners or bilinguals have generally turned out to be deficient monolinguals. One reason given is provided by the lack-of-exposure argument, which can be interpreted as follows: as the bilingual's linguistic experience is split between two languages, s/he is insufficiently exposed to either language to acquire full competence. We will note that this explanation is not compatible with Chomsky's approach as his innatist position reduces the significance of the quality and quantity of language input in language acquisition, and has therefore tended to be ignored. The second and more frequently presented reason is that of interference or (negative) transfer, terms which are frequently considered interchangeable, should however be distinguished clearly as shown in the following discussion.

Subsequently we will discuss the phenomena of interference and transfer in a multilingual context. In our investigation of crosslinguistic contact we will review both the usefulness of Nemser's theory of approximative systems and some of Cummins' hypotheses, such as the BICS/CALP distinction, the common underlying proficiency model of bilingualism and the threshold hypothesis in an attempt to explain some (contradictory) phenomena observed in bilingual or multilingual speakers. Whilst emphasising the shortcomings of the contrastive analysis hypothesis, we shall finally argue for a systems-theoretic approach to transfer research at the end of the next chapter.

Interference has been defined by Weinreich (1953: 1) as 'those instances of deviation from the norms of either language which occur in the speech of bilinguals as a result of their familiarity with more than one language.' It is, however, generally taken to mean the transfer of structures characteristic of L1 to L2 and is distinguished from conscious bilingual transfer procedures such as borrowing and codeswitching.

As argued in more detail later, it would be more precise to speak of transfer when describing this phenomenon and to reserve the term interference for a phenomenon the contrastive analysis hypothesis has shown to exist, that is one language system interfering with the other without resulting in the transfer of structures. Whilst interference also occurs between two equally developed language systems, for instance in ambilinguals, the phenomenon of transfer is obviously more likely to take effect the less balanced the two language systems are. The frequency of transfer will then be proportional to the dominance of one language system over the other, whereas in the evaluation of balanced bilinguals nothing is said about the overall competence of the speaker.

As mentioned earlier, a phenomenon resembling transfer generally occurs at an early stage of bilingual language development in children, when the two language systems are not yet structured and differentiated and widespread mixing tends to occur. As the existence of two or more separate language systems is one presupposition of transfer, to describe this phenomenon as transfer would undoubtedly be a misapplication of the term.

On the basis of the evidence given we may therefore assume that the coexistence (or perhaps even competition) of two language systems is likely to result in unidirectional transfer (in case of one dominant linguistic system), which will lead to a reduction of language competence measured in monolingual terms. We can furthermore assume that the two language systems are likely to interact with each other leading to largely unpredictable results or deviant structures not related to the structures of either language. It is this effect that should be termed interference. As there is ample evidence of crosslinguistic contact occurring in bilingual speakers and second language learners we will in the following draw on some theories developed in SLA research to explain this phenomenon.

Contrastive analysis hypothesis

The traditional contrastive analysis hypothesis (henceforth CAH) claims that the likelihood of transfer occurring can be related to the structural similarity of the two language systems involved: the more similar the two linguistic systems are, the more likely positive transfer is to occur; the greater the difference between the two systems, the more likely negative transfer is to be observed (see, e.g. Larsen-Freeman & Long, 1991: 53). The epithets 'negative' and 'positive' specifying transfer may appear rather confusing in a bilingual context, as these distinctions obviously refer to the facilitation of the language learning process and not to the outcome, as any transfer which is recognisable in the target language system as such is obviously negative as it does not belong to the target language system (see Chapter 3).

Although there appears to be a general consensus of opinion that CAH has not been borne out by the facts, it should not be rejected out of hand. One problem with CAH is that it is most definitely out of fashion as it formed part of a behaviourist interpretation of language learning as habit formation which did not survive the psycholinguistic turn. But there is no reason why we should not be able to assess or apply CAH in a different theoretical context as shown, for instance, by James (e.g. 1992, 1998).

The second problem is that the basic tenet of CAH is theoretically flawed as the pair similarity – difference can lead to unpredictable results. As observed by Osgood (1953: 21), greater similarity between two systems can lead to lack of differentiation and increased (negative) transfer as the systems – though confusingly similar – do not wholly agree. Thus a revised CAH would have to predict that the more similar the two language systems involved are the greater the likelihood of positive or negative transfer. It has also been noted by Hoffmann (1991) and Selinker (1992) that the phenomenon of transfer must be perceived as process-oriented and therefore CAH would have to be complemented by a psychological theory explaining what leads language learners to identify certain target language structures with their native language 'equivalents.'

If research motivated by CAH has failed to confirm the hypothesis, it has certainly established beyond doubt that coexisting language systems do not exist peacefully side by side in ideal modularity, an image Universal

Grammar linguistics would suggest, but that these systems tend to interfere with each other and perhaps even compete (for further details see later). In this respect we can, for instance, assume that, for example, bilinguals do not differ from second language learners.

The idea of interlanguage

In 1972 Selinker introduced the concept of interlanguage to refer to the transitional stage in the SLA process, that is the systematic behaviour of learners of a second, non-native language. It is a theoretical construct still used in SLA research to identify the stages of development in language learners on their way to native language proficiency in the target language.

Nemser (1971), who is usually quoted together with Selinker (1972) and Corder (1971), as the creator of the concept of interlanguage, actually holds a rather more radical hypothesis in SLA research, that is the theory of approximative systems. He claims – in somewhat simplified terms – that at least some second language learners never actually learn their target language, but that these learners at some stage stop developing their foreign language skills and settle for what they have got. Thus when Nemser (1971: 117) states that 'learner systems are by definition transient' this should not be taken to contradict his subsequent statement that 'stable varieties of L_a [approximative systems] are found in immigrant speech, that is, the speech of long-time users of L_T [target language] who, often having obtained considerable fluency in this language, have obviously reached a plateau in their learning.'

This phenomenon, which we refer to as partial achievement, is in part attributed to an effect termed fossilisation, a description of a lack of progress in language learning comparable to the failure of children to learn certain structures or to the development of utility systems, that is, restricted codes used for specific purposes, which is primarily a sociolinguistic phenomenon.

Although the term interlanguage, as developed by Selinker (1972; 1992), has been widely used as a cover term for incomplete or transitional processes of language learning, we nevertheless want to draw on Nemser's conceptualisation of learner systems as our approach to multilingualism focuses on the (dynamic) systems of language learners. Although the differences between Nemser's approach and Selinker's conception of interlanguage are frequently ignored it is Selinker himself (1992: 24–25), who explicitly draws attention to these differences:

> Another introductory point of importance is that the terms 'interlanguage', 'transitional competence' (Corder 1971, 1981) and 'approximate systems' (Nemser 1971) are not synonymous and should

not be treated as such. In my view, they reflect different theoretical positions that have practical ramifications. The transitional competence hypothesis emphasises the in-flux phenomenon of only certain Ils [interlanguages]. This hypothesis does not pretend to account for those Ils which are permanently fossilised or even for the real possibility of those parts of developing Ils which may be fossilised relative to particular contexts. The approximate systems hypothesis is different from the other two in its emphasis on the directionality towards the TL. The latter hypothesis is, I believe, fundamentally false in its view that SLA evolves in stages which gradually more closely approximate the TL. It is in fact a denial of the strong possibility of the reality of permanent fossilisation.

On an individual level there are therefore at least two important consequences to be derived from Nemser's theory of approximative systems: the first being that linguistic systems can be interpreted as transient stages, even if the – necessarily idealised – target language is never attained; the second consequence which – although never drawn by Nemser – appears to us to be implied by his theory is that we can assume that both native and non-native speakers work with approximate systems of a target language. We can therefore suppose that multilinguals also work with approximative systems of their target languages, which implies that their competence in one or more languages is likely to be restricted. Please note that this assumption represents a generalisation of approximative systems theory.

Furthermore we would like to point out in this discussion that classifications of language proficiency such as ideal speaker proficiency and rudimentary speaker proficiency as found in second language learners should be seen as ranges rather than levels in the strict sense of the word. In Figure 1 the shaded areas are supposed to indicate the gradual transition expected from one state to the next.

Summarising, we can state that there is ample evidence that bilinguals do not generally achieve the same levels of competence as monolingual speakers – the exact nature of the linguistic skills in which the bilinguals appear to be deficient will have to be specified though – and that SLA provides us with a number of explanatory hypotheses concerning the reasons for this deficiency.

The conclusion we can draw so far takes an unsurprising form: monolinguals speak one language very well and bilinguals tend to speak two languages, but these somewhat less well. This might also tie up with a theory of general cognitive effort suggesting that, if the effort required to master a language is split between two, it is likely to result in a reduction of the mastery of both.

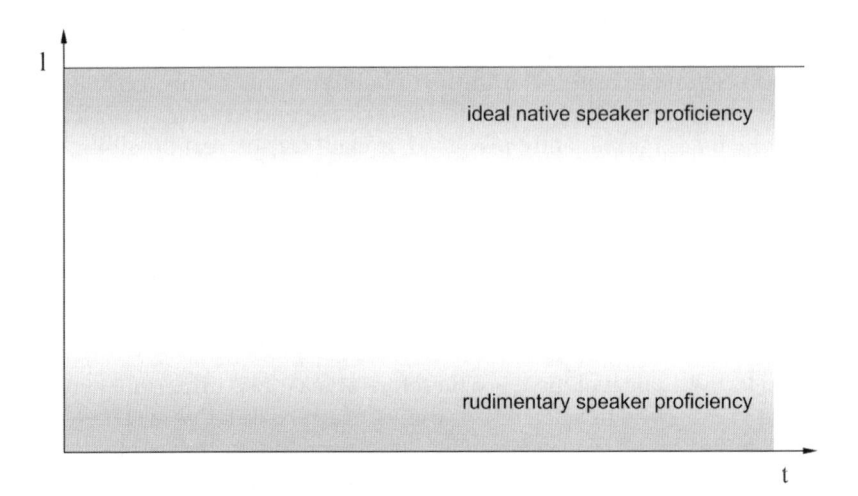

Figure 1 Language range graph
ISP = ideal native speaker proficieny; RSP = rudimentary speaker proficiency; t = time; l = language level

The Impact of the Peal & Lambert Study

In their seminal investigation published 1962 Peal and Lambert found a positive relationship between bilingualism and intelligence in ten-year-old, middle-class French-Canadian bilinguals. French-English bilingual children performed significantly better than French monolinguals on both the verbal and the non-verbal measures in either language. The scholars' conclusions include the assumption of a positive transfer between a bilingual's two languages, facilitating the development of a cognitive system, that is verbal intelligence. This phenomenon of positive transfer, which interestingly enough, appears to take place from L2 to the dominant language, does not only refer to linguistic competence but seems to affect cognitive skills as well. Thus we suggested that this phenomenon should be referred to as the paradox of transfer (Jessner & Herdina, 1994) which we will have to pay particular attention to in a more detailed discussion of language contact phenomena in the next chapter.

The Peal & Lambert study had an enormous impact on research on bilingualism mainly because of the methodology used in their experiments to ensure the validity and reliability of their results and the positive consequences for Canadian bilingual educational policies. Furthermore, the study provoked research interest in factors other than intelligence (see Reynolds, 1991b: 147).

A survey of the investigations reporting the nature of positive consequences of bilingualism resulting from research since 1962 gives a rather heterogeneous collection of findings mainly on the level of higher creativity and reorganisation of information. According to the list of studies in Hamers and Blanc (1989: 50) bilinguals are reported to outperform monolinguals in the following cognitive functions: reconstruction of a perceptual situation, verbal and non-verbal intelligence, verbal originality, verbal divergence, semantic relations, Piagetian concept formation, creative thinking, non-verbal perceptual tasks, verbal transformation and symbol substitution, and various metalinguistic types of performance (see Chapter 6).

Various, nowadays classic, ideas on how to find an explanation for the contradictory results, that is from studies of minority children taught in the majority language whose test performance did not result in an advantage over the monolingual children, were developed by Lambert on social perspectives and Cummins on educational and cognitive perspectives of bilingualism.

Lambert: additive and subtractive bilingualism

Lambert's distinction between additive and subtractive bilingualism (1977) has contributed to the explanation of why investigations of certain groups of children produced mainly negative results, whereas others did not. This sociolinguistic approach to bilingualism represents an important development in bilingualism research as it indicates a move away from a dichotomous classification of bilingualism and monolingualism toward a more complex taxonomy including bilingual subtypes along with the fundamental bilingual/monolingual distinction. The distinction between additive and subtractive bilingualism thus provides the basic ingredients for an explanation of individual and societal bilingualism, which have to be seen as linked phenomena (see Jessner, 1995: 175).

According to Lambert additive bilingualism is used to refer to the positive outcomes of being bilingual and is characterised by the acquisition of two socially prestigious languages. Subtractive bilingualism hence refers to the negative affective and cognitive effects of bilingualism (e.g. where both languages are underdeveloped). This situation occurs when the acquisition of one language threatens to replace or dominate the other language (in members of ethnic minority groups), a phenomenon which is accompanied by subtractive biculturalism. In more recent research Landry, Allard and Théberge (1991) have suggested a wider use of Lambert's concept referring to ethnolinguistic vitality on a social level.

Cummins: BICS/CALP, common underlying proficiency and threshold hypothesis

As mentioned above, the Peal & Lambert study has had enormous consequences on bilingualism research as the results of the study can be considered unexpected in two ways: firstly, we have evidence of something like positive transfer occurring, which seems to have suffered a peculiar reversal as the transfer appears to be taking place from the secondary language to the dominant or primary language. Secondly, this positive transfer does not appear to be restricted to language competence, but affects general cognitive skills as well, something which according to Universal Grammar is not possible, as crossmodular effects are not admissible (see Timm, 1975; see Chapters 4, 8).

Let us first look at possible explanations of this phenomenon, before turning to the consequences. One point to be made is that the example given does evidently not represent a freak result as there is enough further evidence to substantiate the superiority of bilinguals in cognitive tasks (see Baker & Prys Jones, 1998: 62–73; Chapter 5). Not all these findings can be attributed to methodological inadequacies in the research conducted, although research on bilingualism is confronted with some particularly complex empirical problems (see, e.g. Reynolds, 1991b).

A more likely explanation is that the objects of investigation do not coincide or, in other words, that they are not homogeneous as seen by Cummins (1979) who suggested that language competence – as investigated in bilingualism research – in fact consists of two separate skill types or competence modules: basic interpersonal communication skills (BICS) and cognitive/academic language proficiency (CALP), which he later labelled conversational versus academic language proficiency (see Cummins, 1991a: 71). BICS are taken to cover those visible features which are relatively easy to measure, for example, pronunciation, vocabulary, grammar, fluency, etc. whilst CALP, also referred to as cognitive linguistic competence, is taken to denote the ability to make use of the cognitive functions of the language, that is to use language effectively as an instrument of thought and represent cognitive operations by means of language.

This distinction appears to explain the discrepancy between obvious linguistic competence and the lack of conceptual linguistic knowledge observed in some bilinguals in the wake of the semilingualism debate (see Hoffmann, 1991: 127). Cummins' hypothesis has been criticised by several researchers. According to Romaine (1989: 239–240), for instance, this bifurcation of language competence appears to stand on rather shaky empirical footing as subsequent research has not been able to confirm the existence of two separate types of language competence, that is an even partially modular structure. In a response to his critics Cummins (2000), however, presents linguistic evidence and other academic support for the conversational/academic distinction.

The traditional view of bilingualism, as described as double monolingualism before and also known as balance theory (Baker, 1996: 145), suggests that learning two languages, even if simultaneously, is to be seen as a reduplication of the processes required to learn one language and that the command of two languages requires two separate underlying proficiencies (henceforth SUP). From this assumption it is not a great step to the interpretation of human language capacity as a finite quantity, implying that the command of two languages will therefore simply require a splitting up of the cognitive resources available for language processing resulting in semilingualism (see Romaine, 1989: 235).

As an alternative Cummins (1984, 1991a,b) suggests a common underlying (language) proficiency (henceforth CUP) in bilinguals, which can be affected by the use of either language, that is the linguistic development of L2 can show its positive effects in L1. In his model Cummins does, however, not specify which parameters determine the size and development of the linguistic think tank. In a study aimed at finding empirical evidence for the linguistic interdependence hypothesis Verhoeven (1994) found that the transfer as suggested in the model was very limited on the levels of lexicon and syntax but that pragmatic, phonological and literacy skills were transferred as suggested by Cummins. Nevertheless it is highly significant that Cummins was one of the first scholars to offer a number of coherent explanations of the contradictory findings. For one he suggests that language competence is a much more complex phenomenon than originally assumed.

Whilst the bicompetence model of BICS and CALP has been of importance mainly for pedagogical reasons (see Baker, 1996: 151–161; Cline & Frederickson, 1996), CUP represents a far more interesting hypothesis for the development of our approach to multilingualism. The CUP hypothesis represents a major step forward in the investigation of multilingualism in so far as it constitutes a theoretical recognition of the fact that multilingualism leads to the development of proficiencies not to be found in monolingual speakers. In our view this represents a major advance in multilingualism research. It is, however, to be noted that the CUP hypothesis is still too strongly determined by the controversy initiated by CAH, which means that Cummins' understanding of CUP still seems too strongly restricted by the static view of the language systems suggested by CAH. Yet Cummins does stress the importance of the investigation of developmental issues in second language learning (see Cummins, 1991a: 13).

Influenced by Skutnabb-Kangas (1976, 1984), Cummins (1976, 1984) suggests a theoretical framework assigning a central role to the developmental interrelations between L1 and L2 and to the level of competence in both languages. He introduces a threshold hypothesis to explain the apparently contradictory results in studies on bilingual children. He proposes threshold levels of linguistic proficiency which bilingual children must attain both in order to avoid cognitive deficits and allow the potentially beneficial aspects of becoming bilingual to influence cognition.

According to the threshold hypothesis the effects of the interrelation between the two languages can be explained by two proposed thresholds in the levels of the bilingual's competence. At a low level of competence negative consequences of bilingualism cannot be avoided. At a second level – between the two thresholds – with age-appropriate proficiency in one but not the two languages, neither positive nor negative effects are probable, whilst at a third level – beyond the upper threshold – positive cognitive effects are likely to result in a situation that is referred to as balanced bilingualism.

Although the theory has practical limitations (see Baker & Prys-Jones, 1998: 775) Cummins' introduction of the threshold hypothesis represents a revolutionary step in theoretical and applied linguistics. Whilst so far the only commonly applied threshold was that of the critical age hypothesis (see, e.g. Lenneberg, 1967), Cummins introduced a novel scalar measure or intensity criterion, relating to the level of competence acquired by the bilingual speaker. This introduction, which can be interpreted as a qualitative change in the bilingual system, will be turned to again in the detailed description of DMM in Chapter 7.

In conclusion to this chapter we can say that although the double monolingualism hypothesis has been used as the prevailing concept in studies on bilingualism and SLA, there have been several attempts to explain contradictory results in several strands of research. Furthermore the contact between two language systems has not only to be seen as unidirectional from L1 to L2, as described in early SLA research, but as a multifaceted phenomenon in need of particular attention in multilingualism research, which has to be based both on research of bilingualism and SLA.

Chapter 3
Transfer Reconsidered

In DMM we recognise transfer phenomena as significant features in multilingual systems. Both the occurrence of transfer phenomena in multilingual speakers, their attempts to suppress these in a monolingual environment and their conscious use of transfer strategies in language learning constitute prime objects of multilingual investigation. It will, therefore, be necessary to specify some working hypotheses relating to transfer and language mixing phenomena, before turning to other significant aspects of the multilingual system.

- The multilingual system is not reducible to multiple monolingualism

In DMM the multiple language speaker and her/her language system is not merely the result of adding the two or more language systems but a complex dynamic system with its own parameters, which are not to be found in the monolingual speaker. This irreducibility hypothesis runs counter to the attempts of Universal Grammar to provide a unitary theory of language acquisition based on the universal applicability of Universal Grammar principles and/or the principles and parameters approach, which will be dealt with in Chapter 4 in more detail.

- Transfer phenomena should be viewed as a coherent set of phenomena

In DMM we assume that the traditional distinction between codeswitching in research on bilingualism and transfer in SLA research is historically understandable but methodologically unfounded, impeding research on transfer in multiple language systems. In principle, transfer phenomena should be viewed and investigated as a single factor: a differentiation should therefore be made according to systematic criteria and not according to whether they occur in the context of SLA or multiple language processing.

In the tradition of language research, interaction phenomena between one or more language systems are seen as separate. Thus transfer, interference and codeswitching are not considered as individual variants of the same phenomenon, but as separate phenomena which are to be attributed to completely different and unrelated fields of research. Using a systems-theoretic approach, we wish to introduce a new view of interlingual phenomena and to show how the existence of earlier language systems can affect later language systems.

Transfer in Research on Bilingualism

Switch theory and double monolingualism

The tradition of the investigation of codeswitching starts with Weinreich (1953), who refers to the fact that, when a speaker has command of more than one language, both language systems do not coexist as two entirely separate spheres but that a large number of transfer and interference phenomena are to be expected in multilingual speakers.

Apart from objecting to multilingualism as a negative influence on the intellectual and linguistic development of the child, the original approach to the problem of bilingualism lay in an attempt to transfer the insights gained from the investigation of first language acquisition (henceforth FLA) to SLA. It was based on the assumption that (1) SLA obeys the same principles as FLA and that (2) the bilingual or multilingual individual represents a duplication of the monolingual speaker (see Keatley, 1992: 52 and Chapter 2).

This view corresponds to the perception of the bilingual or multilingual speaker as a monolingual speaker in more than one language community. In recent language acquisition theory this hypothesis is also described as identity theory: 'In its extreme form, the identity hypothesis asserts that it is irrelevant for language acquisition whether or not any other language has been learned before; in other words, first and second language learning is basically one and the same process governed by the same laws.' (Klein, 1986: 23)

Starting from the double monolingualism hypothesis, which also assumes that the two (or more) language systems are basically separate, it is not particularly surprising that research on multilingualism initially focused on the interaction phenomena between these language systems assumed to be basically identical. In the course of these investigations a fundamental distinction was made between two types of phenomena: transfer phenomena on the one hand – which in negative terms are seen as interference – and codeswitching on the other hand, which is seen as a (conscious) alternation between two or more language systems.

As far as we know, no research has been conducted according to which the two phenomena are seen as comparable interaction phenomena within two language systems although the distinction has similarly been criticised by Gardner-Chloros (1995: 86) who suggests that '[c]odeswitching should be viewed as an analytical construct rather than as an observable fact. It is a product of our conceptualisations about language contact and language mixing, and it is not separable, either ideologically or in practice, from borrowing, interference or pidginisation.' Lüdi (1996: 242) makes an attempt to explain the difference by introducing two perspectives, that is exolingual and bilingual perspectives, and during his discussion he points out that codeswitching can occur in language learners and compensatory strategies can also be found in competent mono- and bilingual speakers.

In contrast to research tradition, we assume that, from a systems-theoretic point of view, these features can reasonably be viewed as structurally identical. The topic of codeswitching allows us to deal with numerous language mixing phenomena including extrasyntactic codeswitching, intersyntactic codeswitching, borrowing, codemixing and many related aspects whereas research on transfer phenomena has been an issue of SLA research as already pointed out in the previous chapter. Therefore in compliance with linguistic research tradition we intend to begin by discussing transfer and codeswitching as separate features.

Codeswitching and the continuous monitor

Traditional research on codeswitching (henceforth CS) was determined by identity-theoretic assumptions. Early research supposed that the alternation between the use of the two language systems must be governed by some kind of external switch or monitor. One of the suggestions made concerned the existence of a neurophysiological switch determining the alternation of the accessibility of the two language systems ensuring the avoidance of general interference phenomena occurring between the two language systems. This switch theory, as suggested by Penfield and Roberts (1959), soon proved to lack empirical backing. Subsequently Albert and Obler (1978) suggested a continuous operating monitor system governing the switch from the use of one language system to the other (see Keatley, 1992: 18–20). The research initiated by these hypotheses necessarily led to a closer investigation of the actual detailed process of CS, which established itself as a field of research in its own right.

We must, however, bear in mind that CS as such has widespread implications as outlined by Hamers and Blanc (1989: 85):

> This single-switch hypothesis implies the existence of two psycholinguistic systems, one for each language, and a certain degree of independence between two sets of language-specific information processors. The existence of language-specific processors versus a

common mechanism is the major debate in psycholinguistic research on bilinguals.

Codeswitching as alternating between languages

In most research concerning CS one implicit assumption is that CS is basically a performance-oriented phenomenon that does not affect the language systems involved: 'Le point de départ de l'étude est dans la réalité de tout corpus où est repérable le code switching. C'est un corpus de *performances*. Il n'est saisissable que dans la parole concrètement prise.' (The point of departure of the study is the real corpus in which codeswitching is detectable. It's a corpus of performances. It's only comprehensible in recorded discourse. Translation by UJ) (Lafont, 1990: 71)

The interpretation of CS as a purely performance-based effect implies that CS cannot influence the structure of the language systems involved. The attempt to view the language systems in isolation and therefore to suppose they are unaffected is reflected in the assumption that the codeswitches are ordered and thus the universally valid rule-governed nature of natural languages is not endangered. This view was expressed by Weinreich at an early stage of psycholinguistic research: 'The ideal bilingual switches from one language to the other according to the appropriate changes in the speech situation (interlocutors, topics etc.), but not in an unchanged speech situation, and certainly not within a single sentence.' (1953: 73)

The hypothesis that CS phenomena are restricted to the extrasyntactic sphere was soon disproved by empirical evidence provided by multilingual language use (see Bhatia & Ritchie, 1996). The problem of 'orderly transitions' from one language system to the next is obviously a rather more crucial one in Universal Grammar, as the (possible) existence of wild grammars, that is grammars not corresponding to the language-specific rules or the universal rules laid down by Universal Grammar, would threaten the whole concept of Universal Grammar (see Chapter 4). The acknowledgement of the existence of innersyntactic codeswitches resulted in the attempt to determine and specify the conditions of an acceptable switch between two systems. It was assumed that CS had to underlie specific constraints governing and restricting the seemingly erratic switch from one language system to the other.

Thus Pfaff (1979), Clyne (1980), Poplack (1980) as well as Sridhar and Sridhar (1980) suggested internal syntactic constraints, which could, for example, take the form of equivalence, attributive or prepositional constraints and free morpheme constraints. The equivalence constraints presuppose that an essential condition of CS is that L1 and L2 exhibit isomorphic structures in the syntactic surface structures before a switch process can take place. Note that the equivalence constraints are comparable to the transfer conditions specified by CAH.

A further plausible hypothesis is found in the assumption that what is traditionally viewed as an innersyntactic unit, such as a prepositional phrase, epithet or classifier plus noun construction is not likely to be broken by a codeswitch. As argued by Sridhar and Sridhar (1980), Myers Scotton (1990), Pandharipande (1990) and Auer (1995) codeswitch decisions appear to be governed by pragmatic rather than syntactic criteria (see also Heller, 1988) and therefore the assumption of abstract context-independent CS criteria seems to be based on a misconception of how language works. This suspicion is confirmed by the fact that CS can also be observed as an intralexical morpheme-governed phenomenon. As Myers Scotton (1990: 85) states: 'Codeswitching is defined as the use of two or more linguistic varieties in the same conversation. It can be intra- or extra-sentential and also intra-word.' It is also to be observed that CS has been discovered to be a far more varied phenomenon than originally assumed. Baker (1996: 87), for instance, identifies 'thirteen overlapping purposes of codeswitching.' The initial distinction between CS and borrowing, that is between the complete change from one language system to the other (i.e. orderly transition) and the borrowing of individual elements of the embedded language into the matrix language, has turned out to be unsatisfactory as researchers are confronted with nonce borrowings on the one hand and codemixing on the other, without being able to draw a clear line between CS and borrowing.

Thus we can say that both the hypothesis concerning the separability of CS and borrowing phenomena and the hypothesis of the orderly transitions between the respective language systems have been disproved by empirical evidence. Pandharipande (1990), for example, has suggested that codemixing constitutes a transfer phenomenon on a scale of approximation between L1 and L2 and vice versa and thus represents a fusion of the two language systems, casting substantial doubt on any suggestion of the separateness of the language systems or language system processing.

We appear to be confronted with a complex of transfer phenomena which are primarily distinguishable from traditional transfer effects in that they are viewed from the point of view of the interference of (pre)existing language systems and not viewed from the aspect of the acquisition of a second or third language (system) (see Ridley & Singleton, 1995). There is no reason to assume that CS and borrowing are interlanguage phenomena which, compared to transfer, are distinguished by either greater awareness or a substantial difference in the development of either language system. Although it seems obvious that, when dealing with SLA, the two language systems are likely to be unevenly and unequally developed, we are aware of the fact that the language systems in bi- or multilingual speakers are rarely evenly developed either.

Both CS and borrowing can be seen as subject to linguistic competence restraints and situational factors. The distinction between CS, codemixing and borrowing appears to be a grammatical or lexical rather than a psycholinguistic one. This interpretation is applied by Hamers and Blanc (1989: 152) when they state that 'unlike borrowing, which is generally limited to lexical units which are more or less well assimilated, codemixing transfers elements of all linguistic levels and units ranging from a lexical item to a sentence, so that it is not always easy to distinguish codeswitching from code-mixing.' (see also Grosjean, 2001)

On the basis of the given arguments we have to assume that there is no fundamental difference between CS and borrowing phenomena and that they must therefore be interpreted as variants of the same feature which we may term transfer in agreement with traditional terminology.

Transfer in the Learner System

Whilst we may argue that research tradition of CS is relatively short marked by increasing interest where a lot of the research is still to be conducted, transfer – as originally defined – is a phenomenon that has received ample attention. We would therefore hope to find more conclusive evidence concerning the nature of general transfer phenomena in SLA research.

The contrastive approach

Unidirectional influence

As mentioned before, the tradition of contrastive analysis assumed that, where differences between L1 and the language to be acquired were to be found, interference problems would occur, whilst where the languages showed a large amount of structural similarity, L1 would facilitate the acquisition of L2. The language acquisition problems therefore should be attributable to the unidirectional influence of L1. The process concerned is termed language transfer. In case of similarities between the two languages, transfer is assumed to have a positive influence on SLA, and a negative one in case of dissimilarities. This phenomenon could be described as distance theory or theory of structural isomorphy, that is, the success of transfer processes depends on the structural similarities between L1 and L2. Needless to say, this distance-theoretic approach forms one of the theoretical presuppositions of CAH as outlined in the previous chapter.

Bidirectional influence

As has been argued, CAH soon turned out to be too simplistic a conception of transfer, as numerous investigations showed that learners produce mistakes which cannot be attributed to L1. To this problem is to be added the fact that retroactive effects of L1 to L2 could not be excluded. In fact a process termed 'backlash interference' (Jakobovits, 1969) has been observed, which means that transfer is at least potentially bidirectional. This impression is confirmed by Grosjean and Py (1991: 37) who comment on a restructuring effect in L1 due to the influence of L2 in a natural language learning context: 'Et pourtant, il n'est pas difficile de constater que dans une situation de contact prolonge […] la première langue peut être profondément influencée par la deuxième, et ceci à plusieurs niveaux […].' (And yet it is not difficult to notice that in a long-term contact situation […] the first language can be considerably influenced by the second one and this on various levels […] Translation by UJ). More recent research by Franceschini (1999) has shown that the influence between two language systems does not necessarily have to occur from the dominant language but also from the minority on the majority language, in this case from Italian on Swiss-German. In addition to this, further language acquisition and language processing problems were to be observed, which were not attributable to the influence of L1 on L2 and could therefore not be called transfer phenomena in the original, strict sense of the word (as defined above).

As a working hypothesis we might assume that transfer in the above sense is more likely to occur between systems that are essentially unbalanced, whilst interference is more likely to occur between systems that are more balanced. The frequency of transfer phenomena as strict structural transfer is then proportional to the imbalance between the two systems (see Hamers & Blanc, 1989: 87). We may furthermore suppose that the majority of transfer phenomena are more likely not to be reducible to structural influence and are therefore going to fall into the category of unpredictable interference phenomena.

Whilst the principle of linguistic imbalance is confirmed by a number of researchers, Ridley and Singleton (1995), amongst others (e.g. Kellerman & Sharwood Smith, 1986), point out that the perception of language distance is at least an equally important factor. For them it is, however, also clear that researchers must basically assume this influence to be a bidirectional one: 'Where learners perceive a closeness or strong similarity between two second languages, this will typically encourage cross-linguistic influence between them – sometimes, indeed, to the point where such interaction eclipses L1 = L2 influence […].' (Ridley & Singleton, 1995: 125). Ridley and

Singleton also highlight an important aspect of research on crosslinguistic interaction phenomena. It is all too easy to assume that these phenomena occur only in a bilingual context, whilst it is obvious that these phenomena can also be found in L2 and L3 contexts. Although research on multilingualism has to take into account that TLA – for reasons to be outlined in Chapter 5 – cannot be interpreted as a simple repetition of the processes observed in SLA, it will finally have to provide explanations covering the acquisition and coexistence of (at least theoretically) any number of languages.

Crosslinguistic influence

Due to the complexity of interaction phenomena observed between two language systems Kellerman and Sharwood Smith (1986) suggested the theory-neutral concept of crosslinguistic influence (henceforth CLI). CLI was to act as an umbrella term for the effects of transfer, interference and the delayed effects of a change in the factors determining language acquisition. This was taken to include language loss as a change in language competence. It is important to note that CLI is seen as a wider and more flexible concept than that of transfer originally specified.

Transfer is generally observed to occur on all linguistic levels, that is both on a phonological, syntactic, semantic and recently also on a pragmatic level (see Kasper, 1992) – and is occasionally also seen to include orthography. It is also important to emphasise that transfer phenomena have to be seen as related to cultural, social, personal and historical factors (see Odlin, 1989). Beyond this it has been discovered that transfer may not only be observed in speaker errors but also in avoidance techniques, excessive use of certain structures and simplifications (see, e.g. Gass & Selinker, 1994: 88). Transfer is therefore seen as a very complex process and only one among the many determining SLA. The problems of classifying transfer phenomena is last but not least to be attributed to the fact that transfer refers to internal mechanisms which are not necessarily conscious processes. And at the same time they are to be seen as largely unpredictable phenomena in language learning as described by Kellerman (1995). But apart from this we would also like to take other aspects of transfer into consideration.

The Paradox of Transfer

As already mentioned earlier on, the conflicting evidence concerning transfer can be expressed as the paradox of transfer, which is attributed to two factors:

(1) a terminological confusion concerning the type of phenomena to be classified as transfer phenomena
(2) a theoretical confusion relating to the nature of transfer phenomena, which cannot be restricted to specific language modules.

Transfer does not merely take place between the syntactic module of L1 and the syntactic module of L2, but is of an essentially intermodular nature (i.e. the principle of intermodularity of transfer). That is, a large number of transfer phenomena are intermodular and a large number of these phenomena are, strictly speaking, not transfer phenomena at all, but phenomena to be attributed to crosslinguistic interaction to be defined later in our systematic overview of transfer phenomena. This is why we suggested that the term 'transfer' should be used to cover transfer phenomena which result from the application of a structure in one language to a structure in another language. Only in such cases is it useful to speak of transfer and this kind of transfer has to be negative by definition. The use of the epithet positive can only be applied to transfer in a language learning context where the influence of the existence of L1 on the acquisition of L2 can be proved.

As mentioned before, Cummins' Interdependence Hypothesis based on the assumption of a common underlying proficiency (e.g. 1991a) has to be seen in relation to the paradox of transfer as was found in the Peal & Lambert study. His investigation into the influence of L2 on L1 in academic tasks shows that children transfer academic knowledge from L2 learning contexts to their L1. Similarly, in their recent publication on the multifaceted relationship between L1 and L2 Kecskes & Papp (2000) relate the results of their studies to a common underlying conceptual base linking the constantly available sytems which make up the multilingual language processing device to explain the positive influence of foreign language learning on the L1 and vice versa. They further argue that reconceptualisation forms an important part of foreign language learning. The processes as described in their and Cummins' studies are seen as crosslinguistic transfer phenomena and it is also these processes that contribute to our understanding of crosslinguistic interaction in multilinguals. Whereas both suggestions describe a kind of overlap between the two language systems, we would rather see the two languages as two liquids, which, when mixed, acquire properties (such as explosiveness in the case of nitroglycerine) that neither of the liquids had. So these new properties constitute a complete metamorphosis of the substances involved and not merely an overlap between two subsystems.

Transfer Phenomena in Multilingual Systems

In DMM we suggest a distinction between language processing features and systems-relevant phenomena. In contrast to the customary approach of contrastive linguistics we assume that transfer should be seen as a dynamic process and therefore be complemented by a psychological approach explaining what causes a language learner to identify certain structures in the target language as corresponding to L1. This is linked with the concept of psychotypology or perceived language distance as first described by Kellerman and Sharwood Smith (1986).

Although the research motivated by CAH was unable to confirm this, it has become obvious that language systems do not coexist without influencing each other. A particular line of psychological research even assumes that the languages in fact compete with each other (see the competition model by MacWhinney, 1992). In this respect bilingual speakers are indistinguishable from second language learners and, on the basis of what has been said so far, we assume that multilingual speakers are confronted with other systems conditions than monolingual speakers.

It is, however, to be noted that a distinction between preconditions and consequences of multilingualism is difficult to make. Thus it is not quite clear whether parameters such as lateral or creative thinking, metalinguistic awareness or communicative sensibility represent preconditions or consequences of multilingualism (see below). Such results as the fact that multilingual speakers exhibit a better cognitive control of their linguistic processes (Bialystok, 1991b) let us conclude, however, that parameters such as cognitive competence play a greater role in multilingual speakers than in monolingual speakers, where it is commonly assumed that cognitive and language skills can be considered separate. Even if balance theory, as specified by Baker (1996: 145–146), assumes the specific limitedness of the language capacity, that is, that the effort put into L2 occurs at the expense of L1 and vice versa, we must assume that there are natural cognitive and psychological limits to every multilingual system.

A dynamic view assumes that the presence of one or more language systems influences the development of not only the second language but also the development of the overall multilingual system. On the one hand there are psycholinguistic, that is subjective and variable, natural parameters limiting the capacity of the individual to acquire and to command one or more language systems, on the other hand we must also take sociolinguistic factors into account as will be explained in more detail later.

A condition for the development of a plausible and dynamic model lies in the adequate differentiation of those phenomena which are not reducible to the given criteria. In DMM we are required to distinguish between transfer, interference, crosslinguistic influence and crosslinguistic interaction.

The term 'transfer' should be restricted to a basically predictable static or monotonous phenomenon of the transfer of (the same) structures of L1 to L2. The transfer of structures of LS_1 to the learner system LS_2 that has a positive effect on the development of the learner system – as a structurally isomorphous relationship exists between L1 and L2 – can be termed positive transfer. Negative transfer is based on a structural difference between L1 and L2, which means that transfer leads to deviations in the learner system from the expected structures in LS_2. This form of transfer necessarily presupposes an asymmetrical relationship between LS_1 and LS_2.

Interference is intended to refer to those phenomena which are not reducible to either of the language systems involved. We therefore assume that interference is a term to be used to describe language processing, rather than language structure. 'Negative transfer' is rather confusingly also described as 'interference' in the literature on SLA, a concept which should be retained for the description of the result of a dynamic interaction between two or more language systems.

The concept of 'crosslinguistic interaction' (henceforth CLIN) can be taken to include not only transfer and interference as described above but also CS and borrowing phenomena and is thus reserved as an umbrella term for all the existing transfer phenomena. In this sense it is an even wider concept than that of CLI which was originally suggested by Kellerman und Sharwood Smith (1986). These phenomena result from the interaction of two or more language systems. CLIN is also intended to cover another set of phenomena as non-predictable dynamic effects which determine the development of the systems themselves and are particularly observable in multilingualism. Such influences can be interpreted as synergetic and interferential ones. CLIN is not just a category to be added to the existing transfer phenomena but constitutes a significant factor representing the non-reducible dynamic aspect of the multilingual system.

Chapter 4
Universal Grammar Reviewed

It would be impossible to review the development of psycholinguistics and its relevance to multilingualism research without devoting a chapter to the contributions made by Universal Grammar (henceforth UG). Although we do not necessarily agree with the theories, our study would not be complete without a historical and systematic review of the issues raised by UG suggested (see also Brown, Malmkjaer & Williams, 1996). These issues are seen in the context of our endeavour to develop a new understanding of multilingualism. Thus we may note that research interests in the psychological effects of multilingualism presuppose the fairly recent expansion of socio- and psycholinguistics. The question of multilingualism could not have been formulated in the terms of structuralist linguistics alone. As long as language was not clearly placed in the speaker's mind, the question of how this mind manages to contain and process two or more languages could not be put. Thus multilingualism research also rather presupposes the existence of a pycholinguistic turn (see Kasher, 1991) replacing linguistic interest in e-language, that is language perceived as a social fact defined in structuralist terms, by that in i-language as an essentially psycholinguistic phenomenon. This step necessarily results in the development of a theory of competence, which forms a basis of most research on multilingualism as most of the questions asked in this field presuppose some notion of language competence.

We will therefore start this chapter with a look at the competence/performance dichotomy and move on to explain the explicit and implicit beliefs of UG language acquisition theory. The main focus of the chapter will lie on a discussion of the applicability of this research paradigm to SLA and multilingualism.

Theory of Competence

The concept of language competence was originally introduced into mainstream linguistics by Chomsky (1965), who was one of the first linguists to develop an explicit theory of competence. Chomsky's theory of competence is marked by at least two distinctive features: the distinction between native-speaker competence and native-speaker performance, thus allowing for a certain amount of deviation in linguistic data without affecting underlying competence (theory), and the assumption that competence implies a very specific type of knowledge not comparable to other forms of knowledge and not immediately accessible to the native speaker (see, e.g. Brown, 1996).

> Linguistic theory is concerned primarily with an ideal speaker-listener, in a completely homogeneous speech-community, who knows its language perfectly and is unaffected by such grammatically irrelevant conditions as memory limitations, distractions, shifts of attention and interest, and errors (random or characteristic) in applying his knowledge of the language in actual performance. (Chomsky, 1965: 3)

> Linguistic competence is understood as concerned with the tacit knowledge of language structure, that is, knowledge that is commonly not conscious or available for spontaneous report, but necessarily implicit in what the (ideal) speaker-listener can say. The primary task of theory is to provide for an explicit account of such knowledge, especially in relation to the innate structure on which it must depend. It is in terms of such knowledge that one can produce and understand an infinite set of sentences, and that language can be spoke[n] of as 'creative', as energeia. (Chomsky, 1965: 19)

Although we cannot detail the implications of Chomsky's conception here (see Herdina, 1992b; 1996), we might wish to note that it is evidently assumed – though never explicitly stated – that the native speaker and her/his innate faculties are necessarily monolingual. This point appears to have gone unnoticed by many of Chomsky's critics, although it was soon pointed out that the theory of competence proposed was unable to cover the communicative skills required of a native speaker.

Hymes (1972) was to criticise that the model proposed by Chomsky was unable to cover the aspect of appropriacy, which in Hymes' view was to be considered at least as important as grammaticality. He therefore suggested an extension of the theory of competence to include a theory of performance or more systematically a theory of performance competence or proficiency (see Chapter 5).

> In the linguistic theory under discussion, judgements are said to be of two kinds: of grammaticality, with respect to competence, and of acceptability, with respect to performance. Each pair of terms is strictly matched; the critical analysis just given requires that explicit distinctions be made within the notion of 'acceptability' to match the

distinctions of kinds of 'performance', and at the same time, the entire set of terms must be reexamined and recast with respect to the communicative as a whole. (Hymes, 1972: 271)

The broader theory of competence proposed by Hymes intended to include three other aspects identified as feasibility, appropriacy and performance. In the subsequent explication of the notion of communicative competence he introduced the concepts of verbal repertoire, linguistic routines and domains of language behaviour, thus giving the psycholinguistic notion of language competence a stronger sociolinguistic dimension more suited to research on multilingualism. It is to be noted, however, that in his theory of communicative competence Hymes failed to provide a detailed description of the language skills entailed in his conception of language competence. Furthermore, the modifications suggested by Hymes did not imply a criticism of the basic interpretation of language competence as provided by Chomsky (see Hymes, 1972: 281). The fact that language competence might presuppose the existence of other underlying cognitive skills which could or should be included in a comprehensive competence model was not touched upon.

From a Chomskyan point of view linguistic competence can be seen in isolation from general cognitive conditions or sociolinguistic environments. Whilst in Chomsky's case this view is backed up by his innatist position, Hymes failed to challenge or question these more general tenets of UG. We must therefore assume that Hymes did not disagree with the unique nature of linguistic knowledge as suggested by Chomskyan linguistics. This specific uniqueness postulates a language faculty to be viewed in isolation from other faculties, subject of a long lasting controversy with Piaget, as documented in Piattelli-Palmarini (1980), and that general cognitive principles are not applicable to the former. This interpretation, which we might call linguistic isolationism, is complemented by the principle of modularity of mind, which assumes not only that the various faculties of the mind are to be viewed separately, but also that the components contributing to what is termed language competence (i.e. lexical system, syntactic system, phonetic system, etc.) can be interpreted as separate modules as well, although a certain amount of unidirectional input was considered applicable (see Chomsky, 1980: 89).

Tenets of UG Language Acquisition Theory

Despite the fact that UG has established itself as the paradigm in psycholinguistically oriented linguistics, the significance of UG in language acquisition theory was ignored for a surprisingly long time.

Paradoxically this state of affairs was at least in part encouraged by Chomsky himself, who frequently suggested that the insights gained by UG were not to be considered relevant to SLA and corresponding research.

> Let us make a general statement. People who are involved in some practical activity such as teaching languages, translation, or building bridges should probably keep an eye on what's happening in the sciences. But they probably shouldn't take it too seriously because the capacity to carry out practical activities without much conscious awareness of what you're doing is usually far more advanced than scientific knowledge. (Chomsky, 1988: 180)

The last years have, however, witnessed a fundamental change of attitude concerning the applicability of UG to SLA, which can be outlined as follows. Apart from the obvious function of having created a paradigm in theoretical linguistics which had a positive effect both on the discipline as a whole and encouraged abstract theory formation, UG led to a reorientation of linguistics towards the speaker. Whatever one might think of the individual hypotheses of Chomskyan linguistics, the idea of a speaker-oriented linguistic theory rather than systems-oriented linguistics – as, for example, propounded by linguistic structuralism – provides a new theoretical framework in which the problems of language acquisition can be viewed from a new and theoretically consistent angle.

> Although even descriptive adequacy on a large scale is by no means an easy approach, it is crucial for the productive development of linguistic theory that much higher goals than this be pursued. To facilitate the clear formulation of deeper questions, it is useful to consider the abstract problem of constructing an 'acquisition model' for language, that is, a theory of language learning or grammar construction. (Chomsky, 1965: 24–25)

The introduction of a language-acquisition oriented criterion into grammar theory evaluation provided a positive impulse for UG-oriented language acquisition research, and even early theories of grammar were at least expected to develop acquisition-oriented models. The UG model has now been widely accepted and applied as the theoretically most well-founded concept paradigm in language acquisition theory (see, e.g. Felix, 1987; Flynn, 1987; Flynn & O'Neil, 1988; White, 1989).

As there is a great deal of confusion about the tenets of UG concerning (second) language acquisition a certain amount of clarification seems necessary. The positions we believe UG explicitly or implicitly to hold (see e.g. White, 1998) are modularity of mind, a linear theory of language development, a resetting model of SLA, the critical age hypothesis and invariable competence.

It would be incorrect to suggest that UG has not continued to develop its theoretical position since the outlines were drawn by Chomsky at a fairly early stage in theory development (see Sorace, Heycock & Shillcock, 1998). In recent years new positions within the field of UG have been offered such as Optimality Theory (Prince & Smolensky, 1993 quoted from Sorace, Heycock & Shillcock 1998: 5) and researchers have started to find explanations for language phenomena such as attrition, near-native-like proficiency and divergence in outcome in L2 speakers (see e.g. Sorace, 1998; White, 1998). These developments might be judged as positive but also as an attempt to reconcile UG with other research concepts of language acquisition. For our purposes here we consider it important to outline those tenets of UG which have contributed to the distinction of the UG research paradigm from other SLA concepts.

Modularity of mind

Although we are aware of the fact that there is more than one position on modularity as discussed later, we think it necessary to identify a modularist position. Generally speaking psycholinguistic modularism claims that brain functions are domain specific (idiosyncracy of the domain), mandatory (domain specific operations are not subject to choice) and informationally encapsulated (the information processed is only derived from relevant subsystems).

Starting out from the language acquisition device and the independence of language modules, as originally suggested by Chomsky, UG has developed a modular conception of the mind (see Fodor, 1983). Although the modules as such vary in specification, it seems clear that within UG brain functions are neatly compartmentalised, that is placed in (black) boxes responsible for the functions attributed to them. The same principle is applied in the components of UG. Modularity has the advantage of isolating psycholinguistic research from general cognitive science as the thesis suggests that the language module can be treated in isolation from other cognitive functions. It also suggests a separation of general (conscious) cognitive processes and (unconscious) language specific processes (see also Chapter 8).

One possible consequence of this approach is the impermeability or non-interface hypothesis claiming that language learning cannot turn into language competence. Thus Krashen (1981, 1985), for example, claims that learners possess an acquired system and a learned system, both of which are totally separate. The former is developed by means of acquisition, a subconscious process which arises when learners are using language for communication. The latter is the result of learning, the process of paying conscious attention to language in an effort to understand and memorise rules. Weak modularity will admit of a certain amount of interrelation between the modules (see also Chapter 8).

The insistence on the existence of a language acquisition device on the basis of a poverty-of-stimulus argument suggesting that poverty of stimulus makes the assumption of a genetic factor (or language acquisition device) appears a mainstay of UG acquisition theory and in the context of argument Lightfoot (1999: 94) considers it apt to quote Chomsky (1977: 164) suggesting that the UG position has changed little on this theoretical position. Lightfoot's attempt to combine child language, the history of English and evolutionary biology can be seen as an attempt to reconcile a UG-theoretic view of language acquisition by drawing on the insights gained by chaos theory and it is therefore particularly interesting to note which of the arguments have remained essentially unchanged.

We would like to note, however, that the argument of insufficient and linguistically flawed input provides no justification for assuming the existence of a language acquisition device. The presupposition underlying the poverty-of-stimulus argument suggests that learning is an inductive process, where what is to be found in the mind of a speaker has to correspond to the input received from the environment. The spuriousness of this argument becomes clear when one defines learning as result of the interaction between the learner and the environment.

Linear theory of language development

According to this linear model or ladder model language acquisition consists of a number of steps which cannot be skipped and are unilinear, that is, no alternative routes are conceived of. The linear conception ties up with the logical sequence of grammar development for which there is sufficient empirical evidence – if we ignore certain methodological problems involved (see Klein, 1991 on cumulative interpretation). This is stated explicitly by Felix:

> Ignoring details for the moment, the fundamental idea behind this view is that the various principles of universal grammar will successively emerge according to a specific maturational schedule so that at any developmental stage the child's grammar construction will be guided [...] by a proper subset of universal principles. (Felix, 1987: 115)

In our view this linear ladder model (Figure 2) underlies both the

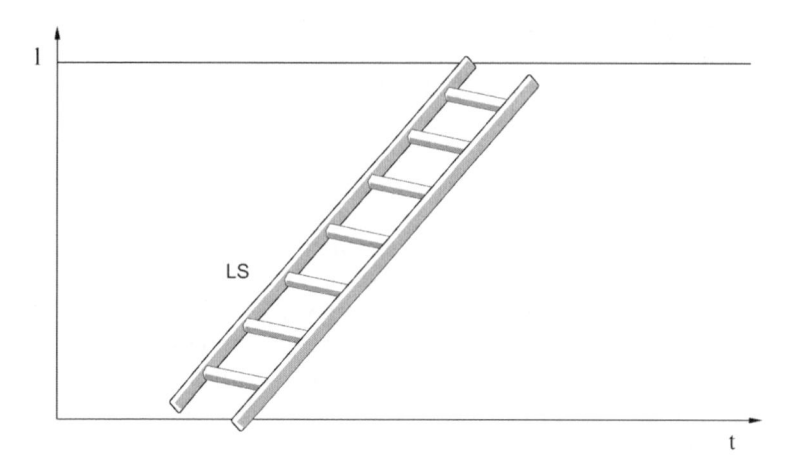

Figure 2 Ladder model according to UG

LS = language system; t = time; l = language level
UG suggests a linear representation of successive steps in the emergence of language competence.

continuity hypothesis (Pinker, 1984) and the maturational hypothesis (Felix, 1987 and Borer & Wexler, 1987).

Resetting model of SLA

According to the principles and parameters approach (first) language acquisition is seen as an experience-based selection of language-specific parameters, which are determined by the principles of UG (see Figure 3). This principles and parameters approach primarily serves the purpose of explaining FLA. On the basis of its popularity as an explanatory model in FLA theory it is generally transferred to the problem of SLA (see White, 1989).

This schematic representation of the development of language competence in the speaker of a language suggests that having passed through a logical sequence of the development of language competence, the speaker achieves a specific invariable level of language competence which is termed the command of a language (see Salkie, 1990: 62).

Building on the logical sequence of language development, UG assumes that second (or third, etc.) language development takes place through the resetting of the parameters set in FLA. This can be envisaged as a ratchet model of multiple language acquisition, as the flexibility introduced is still seen in rather mechanical terms (see Chomsky, 1986: 146). UG encounters a substantial number of problems when dealing with multilingual phenomena, one being the problem of parameter switching or resetting

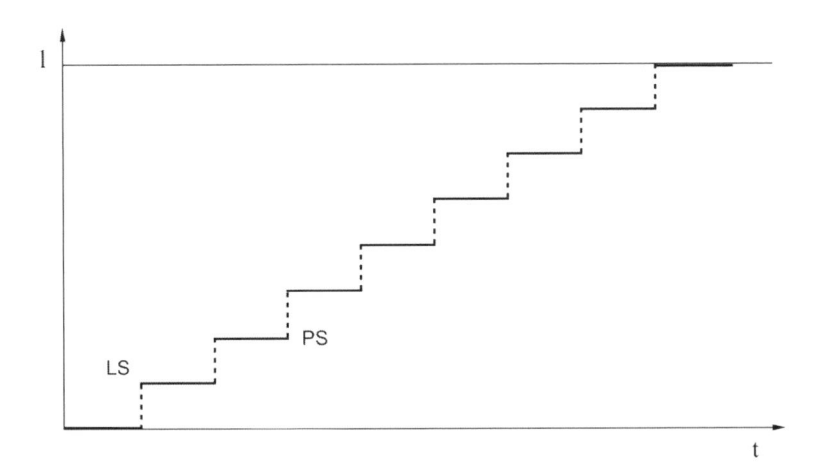

Figure 3 Language acquisition according to the principles and parameters approach
LS = language system; PS = parameter setting; t = time; l = language level
According to UG the language learner acquires a language in accordance with a set of predetermined logical steps.

(between two or more languages) (see Thomas, 1988; Zobl, 1992; Klein, 1995), which is not explained satisfactorily by UG, and another being that of language acquisition between parameters, where the suggestions made by UG, ranging from the innateness of items of the lexicon to the poverty of stimulus argument are rather confusing. See, for example, Chomsky's argument for the innateness of the lexicon:

> [H]uman nature gives us the concept 'climb' for free. That is, the concept 'climb' is just part of the way in which we are able to interpret experience available to us before we even have the experience. That is probably true for most concepts that have words for them in language. We simply learn the label that goes with the preexisting concept. (Chomsky, 1988: 191)

Figure 4 represents an idealised UG interpretation of transitional bilingualism, showing the necessary resetting stages that have to be undergone in the acquisition of the second language. To our knowledge the development of LS_1 is not accounted for. Please compare this model to the illustration of transitional bilingualism in DMM in Figure 27.

The critical age hypothesis

This is not necessarily a part of the UG model of language acquisition but is – or at least was – widely held by UG language acquisition researchers.

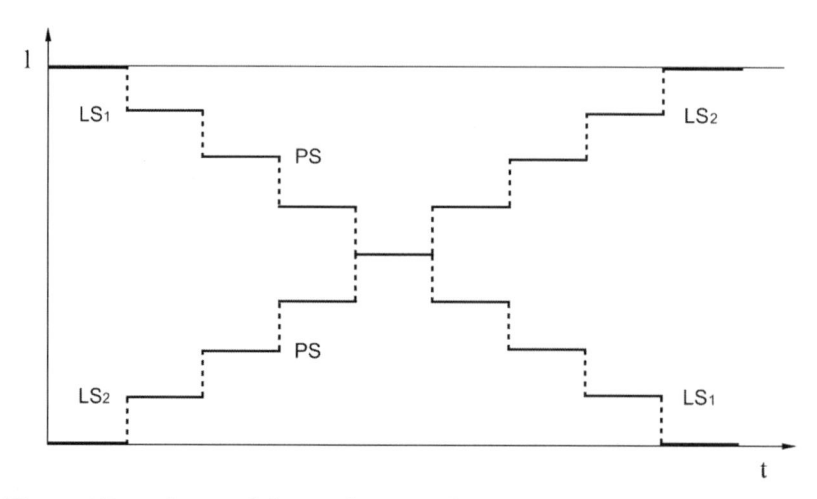

Figure 4 Resetting model according to UG

LS_1= first language system; LS_2 = second language system; PS = parameter setting and resetting stage; t = time; l = language level

The critical age hypothesis suggests that language acquisition has to take place prior to a certain critical age (7–12) and that acquisition after this point in time is likely to be only partially successful, that is, if an individual has not acquired a language prior to a critical age s/he will not achieve native-like competence in the respective language (see, e.g. Penfield & Roberts, 1959; Lenneberg, 1967; see also Singleton, 1989; Singleton & Lengyel, 1995).

This hypothesis is generally backed up by empirical evidence concerning maturational processes going on in the brain leading to an assumed loss of flexibility in the relocation of brain function on cerebral damage. Nevertheless even in the phonetic system which is usually stated as the area where adult SLA is never successful, the strong version of the critical age hypothesis does not stand up to close scrutiny as, for example Bongaerts, Planken and Schils (1995: 43) conclude that 'there appear to be cases of 'late' second language learners who can pass for native speakers phonologically' (see also Palmen, Bongaerts & Schils, 1997). Birdsong (1992) as well as White and Genesee (1996) even reported on some highly successful second language learners who showed native speaker competence in some UG principles.

Apart from the fact that this tenet has thus been disproved in its form, we see it as incompatible with other tenets of the UG model. That is, either UG is not accessible in second language learning, which is incompatible with the idea of a universal grammar, but explains partial attainment, or UG is accessible in second language learning, which conforms to the idea of UG, but leaves empirically observed partial attainment as a problem. Both full attainment and partial attainment are observable in SLA and multilingualism. There is ample evidence of non-native speakers not attaining native speaker proficiency as observed by Huebner (1991: 8), who states:

> [S]econd language acquisition [...] is marked by almost universal failure if success is measured in terms of native-speaker-like intuitions, even among learners who have attained near native-like proficiency in the second language (see Coppetiers, 1987). In the case of bilinguals and second language acquisition, one cannot assume that the final state can be measured in terms of 'idealised' native-like competence.

Although the question of critical age as a problem of accessibility in UG is by no means incontroversial, and the opposition between non-availability of UG in L2 and full availability appears to have shifted towards the idea of attenuated availability of UG in L2 (see later), there appears to be general agreement that after a certain age the nature of language acquisition changes, even if one cannot claim that language acquisition after this age is somehow impaired.

As Figure 5 shows, the core of the critical age hypothesis lies in the assumption that if a certain language level is not attained within a specific age of maturation, the further development of language competence will be adversely affected. It is to be noted that this graph does not show multilingual development, although it is commonly assumed that, for example, FLA can lead to a retardation of second language development in a learning context. LS' and LS" simply express the degree to which this retardation can be postulated to take place.

The widespread adoption of the critical age hypothesis has had a number of consequences in multilingualism research, which need to be explicated. Firstly, the critical age hypothesis suggests a partial or non-accessibility view of SLA (discussed below) thus clearly distinguishing between childhood or primary language acquisition and adult SLA, and accepting the non-transferability of FLA findings to SLA. Secondly, the critical age hypothesis therefore more importantly suggests that the findings of SLA research are not transferable to multilingualism research as, for example, early acquired simultaneous bi- or multilingualism differs from multilingualism achieved by SLA or TLA, etc. This tenet is logically connected to the age-related fundamental difference hypothesis as specified by Bley-Vroman (1989).

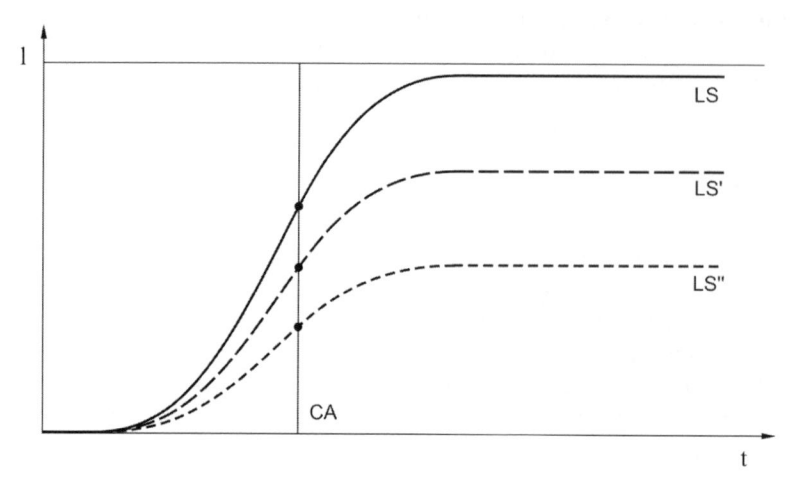

Figure 5 Critical age hypothesis

LS = language development if achievement at critical age is sufficient; LS' = retarded growth due to critical age underachievement; LS" = severely retarded growth due to critical age underachievement; CA = critical age; t = time; l = language level

As has been pointed out, DMM does not adhere to this principle and therefore needs basically not to distinguish between acquired or learned multilingualism. This must, however, not be misunderstood as to mean that maturational effects can be completely ignored. As is well known, maturation has a profound effect on cognitive processes which are most obvious in youth and old age (see, e.g. de Bot & Clyne, 1989; Hyltenstam & Obler, 1989; Hyltenstam & Stroud, 1993).

Invariable competence

As was noted early on in this chapter, one of the consequences of the psycholinguistic turn resulting from the success of UG meant that the central notion of language structure was replaced by the concept of competence (see Chomsky, 1965: 4). Two factors are fundamental to the idea of competence detailed by Chomsky: first, the idea of competence as underlying performance, and second, the idea that this competence is not subject to the variations observed in performance, but represents a constant quality (see, e.g. Sharwood Smith, 1994: 107). This means that competence is seen as a quality that must be considered invariable and finite in the speaker. We can term this the constancy or invariability of competence principle, which in our view is retained despite recent UG-oriented publications on variability in language competence (see, e.g. Towell & Hawkins, 1994; Wardhaugh, 1993: 5).

As within UG this tenet is more of an unquestioned presupposition rather than an explicit hypothesis all the main principles of UG (innateness, principles and parameters, competence versus performance) indicate that competence is assumed to be invariable and would therefore not allow for the possibility of language depreciation or individual loss. Invariable competence is, therefore, incompatible with a theory of dynamic language development.

The more well-known explicit tenet of UG concerning language competence is that of homogeneity. That is, all speakers of English are assumed to have the same command of the same language system. Thus Salkie (1990: 60), for instance, states that variation is assumed to be messy and therefore not a suitable subject for UG – despite one of Chomsky's more recent contributions:

> Every human being speaks a variety of languages. We sometimes call them different styles or different dialects, but they are really different languages, and somehow we know how to use them, one in one place and another in another place. Now each of these different languages involves a different switch setting. […] So the child knows that this is the way you talk to your friends, this is the way you talk to your grand-mother, and so on. (Chomsky, 1988: 188)

A presupposition of both these tenets is the assumption that language competence is unitary (see Chapter 5).

UG on SLA

As outlined above, one of the tenets of UG is the resetting model of SLA which we would like to present in more detail here. We can distinguish three fundamental UG-relevant positions: the first being the full access or dissociative view specifying that UG is available in FLA and SLA; the second being the no-access or fundamental difference view suggesting that SLA follows completely different principles from FLA and finally the partial access or attenuated availability view suggesting that UG can be accessed in a limited way in SLA (see, e.g. Flynn, 1996).

The full access view can also be termed dissociative as it is assumed that SLA replicates FLA and is therefore dissociated from the earlier process. Partial access views are associative in so far as they assume that SLA is not simply a repetition of the processes involved in FLA, but that SLA somehow builds on the experiences associated with FLA. Sharwood Smith

(1994: 160) refers to these contrasting interpretations using the terms fossilised UG hypothesis, recreative hypothesis and resetting hypothesis.

Full access view or dissociative hypothesis

The most obvious explanation provided for SLA is that UG is available in SLA. The theory of the accessibility of UG is, however, confronted with two problems: first as already noted, the theory of the accessibility of UG is incompatible with the empirically evidenced inadequate command of L2 observed so frequently. This problem will be referred to as the partial achievement problem. Accessibility theory cannot explain why L1 competence level frequently differs from L2 competence level. Second, a dissociative model of language acquisition (see the double monolingualism hypothesis as specified above) is not compatible with the idea of interference (as a negative transfer phenomenon) and cannot therefore attribute L1/L2 divergence to interference. Empirical research on bilingualism has, however, provided ample proof of such interferences in multilingual language processing taking place (see Chapters 2, 3). Third, the dissociative hypothesis stands in clear contradiction to the critical period hypothesis mentioned above suggesting that age-specific factors are relevant to the language acquisition process. We might therefore be tempted to give preference to an associative hypothesis in language acquisition theory.

The UG specific dissociative hypothesis corresponds to the fundamental difference hypothesis as specified by Bley-Vroman (1994 quoted by Singleton, 1995: 7), whilst full access theory is, for instance, specified in Ellis (1994: 442–447). These various theoretical approaches share the view that SLA differs substantially from FLA, even if they vary in the terminology used and details of theory interpretation.

No direct access view or associative hypothesis

The associative hypothesis gives up the idea of the separation of the two (or more) language systems and rather assumes an asymmetrical relationship between language system 1 (LS_1) and language system 2 (LS_2) as clearly identifiable components of the language learner's psycholinguistic system. This allows researchers to retain the critical stage hypothesis specified above. The parameters of her/his first language are thus transferred to LS_2 and then adapted accordingly. UG is therefore assumed not to be immediately accessible in SLA. This does have the specific advantage that inadequate SLA or partial attainment is not a core problem for UG as the latter is assumed not to be available or to be only partially available. The resulting disadvantage is obviously that the UG model cannot be drawn upon to explain SLA as UG is not accessible in LS$_2$ (see Clahsen & Muysken, 1986).

The question of non UG-compatible rule formation in SLA thus remains a problem. As long as UG does not develop a systematic theory of interference, the problem of evident parameter deviation in second/multiple language acquisition persists, forcing UG-linguists to assume wild, unruly transitions in the phases of language development (see parameter settings) in order to be able to explain the empirically observable phenomena. Temporarily deviant individual language systems, interpreted as phases of development, are described as wild or unsystematic grammars (see Clahsen & Muysken, 1986; Duplessis *et.al.*, 1987 quoted by Sharwood Smith, 1994: 157–158).

The recognition of the existence of wild grammars represents an admission of failure for UG. The explanatory deficit results from the fact that UG is forced to accept that there are transitional states between LS$_1$ and LS$_2$ which are not predicted by the UG interpretation of L1 or L2 and their stages of development and thus represent a separate field which is most problematic for UG.

The UG paradox

We can therefore determine: if the two language systems LS$_1$ and LS$_2$ are viewed dissociatively, then the accessibility of UG in SLA is incompatible with widespread partial attainment in SLA, that is UG is unable to explain why the level of L2 competence is not equivalent to that of L1 competence. For obvious reasons the interference argument as an explanation of the differences in achievement levels is not applicable to the dissociative concept. The associative hypothesis on the other hand discards the idea of the separateness of both language systems and assumes an asymmetrical relationship between LS$_1$ and LS$_2$. The parameters are transferred from LS$_1$ to LS$_2$ and then adapted accordingly. We can assume that UG is not taken to be effective or accessible in LS$_2$. This does have the (strictly limited) advantage that partial attainment is no longer the theoretical problem it was for UG. On the other hand UG can then not be applied immediately to the explanation of SLA. Whichever of these two UG hypotheses we select, the problem of non UG-conforming rule-formation or wild grammars persists (see also Bates, 1997: 169).

Partial achievement models

In view of this problem it is hardly surprising that within and outside UG new explanatory models have been developed for SLA. The idea of interlanguage presents itself as a possible way out of this dilemma (see

Chapter 2). We will therefore subsequently turn to a number of interpretations and models dealing with the question of partial achievement, suggesting that there must be another level of language competence in between L1 and L2.

We shall briefly discuss the following explanatory models: approximative systems according to Nemser (1971), the concept of interlanguage according to Selinker (1972) and the adaptive control of thought model by Anderson (e.g. 1983). As already mentioned in Chapter 2, the present interlanguage discussion is still largely influenced by Nemser and Selinker. Nemser is, however, frequently only referred to in the context of the history of the field, whilst Selinker's concept of interlanguage has left a more lasting impression and recently has been applied to multilingual speakers (de Angelis & Selinker, 2001; see also later).

Approximative language systems

Originally one might find it rather difficult to understand the amount of attention paid to interlanguage, a construct already outlined in the historical overview provided in Chapter 2. Is it indeed any more useful than the claim that someone crossing the road is to be found on an interpavement? Within the context of argument outlined above, however, the introduction of interlanguage has an important theoretical function. As defined by Selinker and others, interlanguage is supposed to reflect the hypothesis that learner systems are not merely transitional phases but represent systems in their own right, which obey their own rules. The assumed autonomy of learner systems appears to provide a solution to the problem of limited accessibility and partial achievement.

If interlanguage is an autonomous system which is neither reducible to LS_1 nor LS_2, then the problem of the accessibility of UG is not as pressing as it was, as we can assume that interlanguage is not a natural language (see also Adjemian, 1976) and therefore UG principles do not have to apply to it. Interlanguages would accordingly not have to obey the principles of UG, and we have thus found a nice home for our wild grammars.

The second problem interlanguage appears to deal with is that of partial achievement. As interlanguage, the learner system does not ever have to reach the LS_2 stage, but can remain an approximation of L2 which never merges with LS_2. The most frequently cited example here is that of phonological fossilisation. By introducing interlanguage to deal with the problem of partial achievement, learner systems are neatly compartmentalised so as not to interfere with the claims of UG. Although little is said about this, we must assume that the concept of interlanguage can also be applied to utility systems and pidgins or creoles. This identification is suggested by the application of the term learner pidgin to the learner system (see Schumann, 1978, 1986, 1990; Cook, 1993a: 69–78).

If we, however, bear in mind that some learners – although not many – manage to cross the borderline between interlanguage and LS_2, we realise that the concept of interlanguage shifts the problems rather than solves them.

We have already outlined the view of the multilingual speaker as a deficient monolingual speaker. Although we have argued that in research on multilingualism this must be considered misguided, the idea of partial achievement is a most significant and useful one. As noted before, one of the first linguists to address the evident problem of partial achievement was Nemser, whose theory of approximative systems tries to develop a systematic interpretation of the fact that second language learners are observed (1) generally to achieve only partial command of their respective L2 and (2) that these L2 systems appear to obey their own rules, which are reducible neither to those of L1 nor to those of L2 (see Chapter 2). The most pronounced version of partial achievement (variant 1) is observed when L2 learners develop a stable level of command of an L2 which shows no sign of further development and which exhibits systematic deviation from the rules of L2 preventing the interpretation of the learner system as a strict subsystem of L2. This phenomenon is partially attributed to an effect described as fossilisation by Selinker (1972) – although this might simply be taken as an identification of lack of language learning progress observed, comparable to the failure to learn certain structures found in children – and partly to the development of utility systems, that is, restricted codes used for specific purposes, a (socio-)linguistic phenomenon we shall return to later.

The existence of partial systems is a fact that has received increasing attention in recent years. Partial systems must be taken to include both learner systems and utility systems, as the question of partial command of a complete language system or complete command of a partial system is – from the point of view of approximate systems theory – undecidable and therefore irrelevant. This essential indistinguishability of transitional learner systems and fossilised utility systems is reflected in the common use of the term 'learner pidgin' as mentioned above. What is, however, reflected in the term and the corresponding research is the fact that the learner's approximative system must be seen as an autonomous one obeying its own principles. At this state of discussion we propose that multilinguals also work with approximative systems of their target languages as illustrated in Figure 6, which is another way of saying that

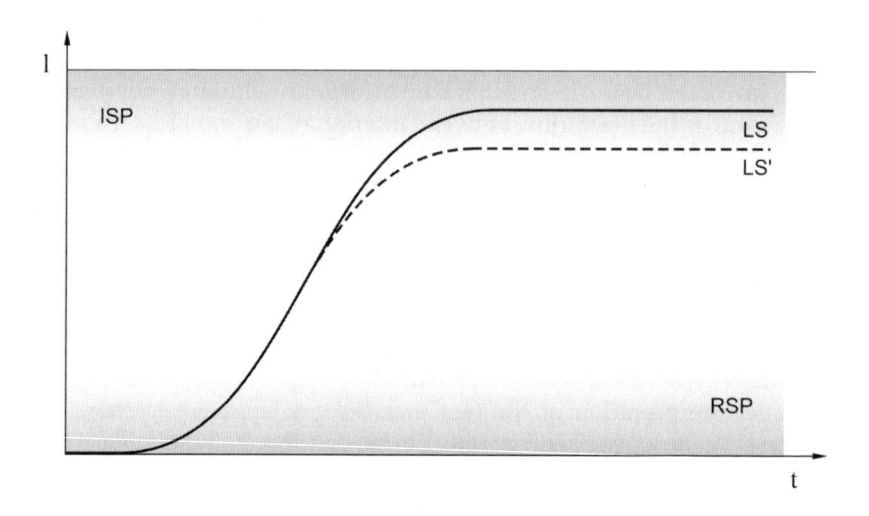

Figure 6 Approximative systems model

LS = language system of native speaker; LS' = language system of a multilingual speaker; ISP = ideal native speaker proficiency; RSP = rudimentary speaker proficiency; t = time; l = language level

their language competence in one or both languages is likely to be restricted.

This graph shows language acquisition according to the theory of approximative systems. Note the flattening of the learning curve prior to the achievement of an absolute, stable language competence (native-like). According to this model the achievement of full language competence does not occur in principle due to fossilisation.

Summarising, we can state that, corresponding to the learner pidginisation hypothesis (Schumann), approximative systems theory (Nemser) acknowledges the fact that learner systems are approximations of standardised conceptions of native-speaker systems. The fact that native-speaker systems are standardised does, however, indicate that native speakers themselves also only have at their disposal a partial or approximate system of LS_1. This appears particularly self-evident, when we consider the plethora of utility systems contained in any developed language system.

Adaptive control of thought

The adaptive control of thought model, generally referred to as ACT model, has recently been turned to by a number of UG theoreticians in the hope of deriving new solutions to the well-known problems. What the ACT model does for UG is to introduce a transitional model of L2 acquisition, which allows for a transition from the learner system to the L2 system via three stages: (1) the declarative or cognitive stage, (2) the associative stage and (3) the autonomous stage. As suggested by Anderson (1983), this represents a process of increasing automatisation through which (in part) explicit learner knowledge is internalised finally to form part of the (UG compatible) L2 system. Obviously this process requires an intermediate stage, which in Towell and Hawkins (1994: 247) is found in a system of internally derived hypotheses representing the learners model of L2. The ACT model runs counter to Krashen's impermeability hypothesis, but seems rather better founded. It also establishes the possible existence of an interim grammar which is related to, but not reducible to L2.

It is to be noted that Anderson's ACT model sits rather uneasily with UG, as Anderson's theoretical outlook blatantly contradicts many of the assumptions of UG, such as the modularity principle, as observed by Cook (1993a: 246):

> A major alternative approach to linguistic theories of acquisition is that of John Anderson, who insists that 'the adult human mind is a unitary construction.' Apart from a few minor adaptations, 'the language faculty is really the whole cognitive system,' evidence for this being the remarkable communalities between languages and other skills.

This move can also be considered an implicit admission of the shortcomings of the UG contribution to the explanation of SLA. It is finally to be seen as an admission of the untenability of the UG presuppositions of the accessibility of UG and full achievement in language acquisition (see also the discussion of modularity and holism in Chapter 8).

UG on Interim Language and Multilingualism

Within the interlanguage discussion in the stricter sense we can distinguish between two concepts: Selinker's (1972, 1992) model of interlanguage – despite claims to the contrary, not a UG-specific concept – and the attempts of UG to use the concept of interlanguage to develop a new approach to the problem (see below). In this context we shall turn to the use of the concept of interlanguage within the framework provided by UG.

The schematic representation in Figure 7 attempts to introduce an understanding of native and non-native language competence levels

[Figure: Intermediate range model showing language level (l) vs time (t), with "ideal native speaker proficiency" at top, "intermediate range" in middle, and "rudimentary speaker proficiency" at bottom]

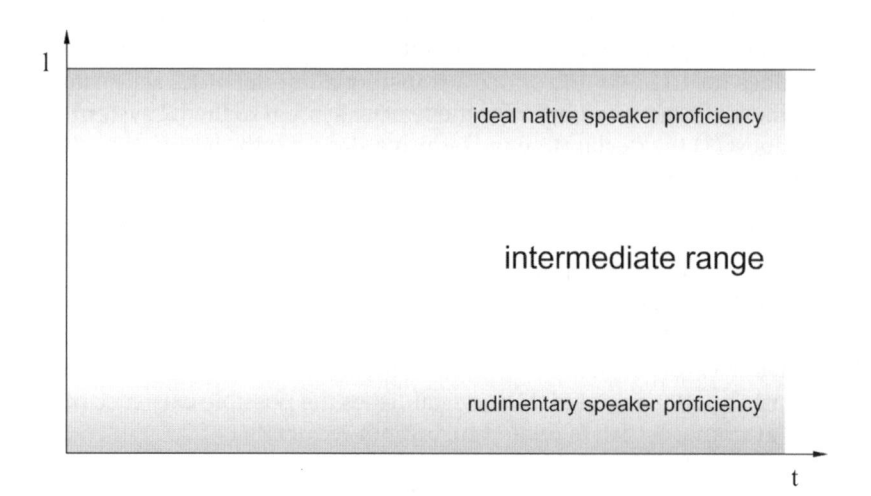

Figure 7 Intermediate range model
t = time; l = language level

admitting of the insights into the existence of interlanguage. We are primarily concerned with the region of rudimentary-speaker proficiency, originally assumed ideal native-speaker proficiency, and an independent level of interlanguage, which is here described as intermediate range.

Here we specify the intermediate range between ideal native speaker proficiency and rudimentary speaker proficiency according to approximative systems theory. Note that the proficiency levels essentially refer to proficiencies obtained in a multilingual context. As a monolingual measure the proficiency measure would have to be replaced by a monolingual construct such as competence. Although we assume that the transition from rudimentary speaker proficiency level to intermediate range is characterised by a number of intermittent stages (see Figure 1), these are not shown here or in subsequent graphs.

As has already been mentioned, the problem of interlanguage is discussed in UG in terms of limited or partial accessibility in L2 (see White, 1989). Furthermore the partial attainment argument acknowledges the existence of interlanguage, which forms a theoretical concession in so far as interlanguages are seen as possible subsets of natural languages. The question of the transition of possible grammars to wild grammars is, however, not answered (see also the problem of language mixing outlined earlier on).

As for the principles and parameters of UG that have so far been investigated, much of the research suggests that L2 learners are restricted by

UG. They do not produce forms that constitute violations of the principles of UG, nor interpret utterances in ways inconsistent with principles of UG. In other words, the IL of the L2 learner is a possible grammar, where a possible grammar is one that falls within the range permitted by UG, even though it is often not equivalent to the actual grammar internalised by native speakers. (White, 1989: 175)

The idea of possible, and in the strict sense, not natural language (developed in response to Adjemian, 1976) represents an extension of the original concept. It is, however, to be doubted whether the modification is sufficient to make UG a plausible explanatory model in SLA. As Selinker (1996: 108–109) points out himself, by drawing on his earlier work in 1992, in forming interlanguage competence, interlingual identifications are the basic learning strategy, with the learner (in ways we do not understand) trying to make the same what in systemic terms cannot be the same. When the learner perceives that the basic learning strategy does not work, then and only then, do UG principles, and learnability click in.

The appropriacy or adequacy of the UG approach is also doubted by Cook (e.g. 1991, 1993b, 1996). He draws both on language acquisition and competence-based arguments to prove the inadequacy of UG in this field. In one of his first critical articles on this topic Cook (1991) questions one of the dogmas of UG, first voiced by Chomsky in his comments on the ideal native speaker, that is the finite state concept of language competence, which can also be interpreted as a uniformity requirement: as there exists a language system which is internalised by all the speakers of a language this language consists of a steady state and deviations such as partial attainment have to be attributed to performance (see Chomsky, 1965: 24).

The solution suggested by Cook lies in the introduction of the concept of a wholistic view of multicompetence. Wholistic multicompetence represents a complete emancipation from the UG concept of language acquisition as a wholistic view necessarily sees both the language systems of a multi-/bilingual speaker not as the sum of two individual systems as such. The most significant arguments for the appropriacy of this view, which Cook (1993b: 3–4) – referring to Grosjean (1982, 1985) – suggests are:

- bilingual speakers have a different knowledge of the first language;
- bilingual speakers have a different knowledge of the second language;
- bilingual speakers command a different type of language consciousness;

- the language relevant cognitive processes of bilingual speakers differ from those of the monolingual speaker.

Cook provides a number of arguments supporting the assumption of a unique and different language system in multilingual speakers. Thus he assumes that (1) the language systems of multilingual speakers refer to the same mental lexicon which makes the frequently quoted CS – presupposing the existence of two separate language systems – implausible, and that (2) the switch-theoretical interpretation of multilingual processing (one language is switched off, whilst the other is activated) is refuted by recent research pointing to the co-presence of both language systems and finally, that (3) both language systems are processed in the same hemisphere and thus L1 and L2 language processing cannot be separated on a neurological level (see Cook, 1993a: 4–5).

Furthermore – as was mentioned earlier on – UG oriented discussion of language acquisition has in the past also paid too little attention to the problem of variability in all its facets. Originally it was viewed by UG rather as a distorting factor and not as a necessary element in a model of language competence (see Gregg, 1990). Empirical research, however, shows that variability (not merely on the performance level) is the norm rather than the exception, even if we discount the necessary idealisation in theory formation. It has already been determined that variability can be interpreted in different ways, first as an idiosyncratic deviation from language norms in the individual speaker and second as systematic deviation from a specific language norm by groups of speakers.

Thus research on second language variation has identified two forms of variability: systematic variability, as investigated by Tarone (1983, 1988), and non-systematic variability as observed by Ellis (1985). Interlanguages can be shown to vary in a systematic way depending on the strategies employed by the speaker and the situation in which the language is used, that is, language production will systematically conform with a rule/principle and then systematically fail to conform with a rule/principle depending on the language use situation and the stage of language development. This systematic variation is complemented by non-systematic variation, a phenomenon we later refer to as scatter (see Chapters 6, 7).

These types of socio- and psycholinguistic variability are, however, more easily integrated into a UG model than the problem of variation in time. Yet empirical research again shows that language competence may not only vary among speakers of the same language but also within a speaker as a function of time (see Vogel, 1990: 51–52). Whilst the first set of results forces us to assume that competence in L1 is not a constant but a variable, the second set of results points to the fact that we must expect to find growth and decay of a language system within an individual speaker (see, e.g. Hyltenstam & Viberg, 1993). Whilst these processes do not appear to be very significant in monolingual speakers, such phenomena are very pronounced in multilingual speakers.

In view of some of the problems outlined so far (transfer, interference, partial achievement, individual variability) it should have become clear that language acquisition theory is not simply transferable to SLA and TLA and their corresponding problems. As already argued by Grosjean and Cook, it seems appropriate to adopt a more wholistic and differentiated approach towards multilingualism than current linguistic theory would provide.

Chapter 5
Multilingual Proficiency Reassessed

This chapter is supposed to provide an overview of multilingualism research (henceforth MR) with the aim of establishing the basis for the development of our model of multilingualism in the subsequent two chapters. We shall therefore discuss those findings from research in SLA and bilingualism which are considered relevant for the modelling multilingualism within DMM and for future research in this area. After a definition of multilingualism and multilingual proficiency with an attempt to clarify terminology in use, an overview of current issues in MR will be presented with a special focus on research on TLA and trilingualism. Variability will be focused on again in this chapter to discuss its essential nature in our approach to multilingualism. Sociolinguistic aspects of variation as the dominant field of investigation will be discussed before we move on to psycholinguistic variability with a particular emphasis on language aptitude representing individual factors in language learning. Finally we will summarise the first part of the book by pointing out those ideas favouring a dynamic view of multilingual systems.

Defining Multilingualism

It is important to note that the object of investigation in a systems-theoretic model must necessarily be multilingualism, which, in general terms, can be defined as the command and/or use of two or more languages by the respective speaker. Most of the research conducted so far has focused on a particular variant of multilingualism, that is, its most common form, bilingualism, and it is assumed that most of the findings obtained in bilingualism research can be generalised to cover 2+n languages. In other words, the findings from studies on language attrition and loss, for instance, in the historically determined research area of bilingualism and SLA support our model of multilingualism, within which bilingualism is treated as a variant of multilingualism thus allowing a generalisation of predictions and findings (see, e.g. Haarmann, 1980: 13). The study of the multilingual psycholinguistic system containing more than two languages is on the one hand based on results of bilingualism and second language research, that is it is similar to systems with two languages, but on the other it necessarily also shows differences to psycholinguistic systems containing two languages.

The model to be developed here serves as a link between SLA and bilingualism in so far as it can explain multilingual acquisition patterns which traditional research has not been able to do. For instance, the case of a trilingual speaker who acquires a foreign language whilst growing up in a bilingual family (simultaneous bilingualism), could, according to current research methods, be only inadequately discussed. Within our model we are able to draw connections between learner systems and bilingual systems and consequently deal with the ongoing changes in individual multilingual proficiency.

Terminological clarification

As can be detected from the title of this chapter and the ongoing discussion we have given preference to the term multilingual proficiency over other terminology such as multilingual competence or multicompetence as suggested by Cook (e.g. 1991, 1993b) to describe individual multilingualism. We consider it important to focus on some of the concepts of language knowledge in order to present our view on this salient issue in the study of multilingualism as numerous models of language knowledge and skills, stemming from fields as different as neurolinguistics and language teaching research, have been suggested, bearing different names.

As pointed out by Baker (1996: 5–6), there are a variety of terms that are used in language learning and language testing, such as language ability, language proficiency, language achievement and the dichotomy of competence/performance which either describe the same construct or are regarded as characteristics of the heuristic construct forming the basis of research in that area. Like language ability, language proficiency is often used ambiguously and both terms are often employed as umbrella terms. In general, it has to be noted that there is no standardised use of the terminology discussed.

Language competence

As outlined in Chapter 4, the theory of competence was one of the main contributions made by early Chomskyan linguistics to psycholinguistic research. Yet Chomsky was not only too ready to admit that his conception of competence was strongly idealised but Chomsky's conception of competence was also restricted to the notion of grammatical competence. According to the critics (e.g. Hymes, 1972) this idealised conception of competence requires a substantial extension by other types of language knowledge. A surprisingly late objection to the UG interpretation of grammatical knowledge was thus the fact that UG is essentially a syntax-based rather than a text-based grammar. In the course of the development of grammatical theory it has become increasingly obvious that in the creation of sentences the speaker draws on a wide range of textual knowledge. The notion of grammatical competence has therefore to be extended to include both intrasyntactic and extrasyntactic, that is (con)textual knowledge. To this is necessarily to be added some kind of pragmatic component to obtain an adequate notion of the types of knowledge necessarily involved in language competence (see also Lyons, 1996). In more recent research conceptual fluency including metaphorical competence as a basis for grammatical and communicative knowledge has also been introduced into the discussion (see Kecskes & Papp, 2000).

Even if we accept that language competence is necessarily multicomponential, we are still bound to the interpretation of multicomponential language competence as a unitary skill or ability. At least UG has always taken a unitary approach to language competence, which itself is based on its separatist or modular interpretation of language functions, that is language functions are defined as being autonomous and governed by such innate components as the language acquisition device. An obvious implication of these assumptions is that there must be some clearly definable entity corresponding to language competence.

Brown (1996: 3) comments on Chomsky's concept of competence:

> Chomsky further elucidates the notion that he wishes to identify by the term 'competence': 'The term 'competence' entered the technical literature in an effort to avoid the slew of problems relating to 'knowledge', but it is misleading in that it suggests ‚ability' – an association I would like to sever. (Chomsky, 1980: 59). Chomsky here draws a distinction between (1) knowing (the forms of) a language, (2) the ability to use the language that one knows and (3) actually using it.

Thus Chomsky's statement might be seen in contrast to Bachman (1991) who describes language ability as consisting of language knowledge. Bachman views it as sometimes referred to as language competence, and cognitive processes, or procedures, that implement this knowledge in language use. This approach is also consistent with information-processing, or cognitive models of mental abilities, which also distinguish processes or heuristics from domains of knowledge (see Sternberg, 1988). Language use then involves the integration of multiple components and processes.

Language proficiency

An overview of proficiency research allows us to identify the following trends. Most researchers (e.g. Harley *et al.*, 1990b; de Jong & Verhoeven, 1992) concerned with the history of research on language proficiency report two phases: the first lasts from the 1920s to the 1980s and is characterised by factor-analytic work or psychometrics. The most well-known contribution was made by Oller in 1976 who introduced a unitary competence hypothesis stating that language proficiency consists of a general factor and is therefore seen as a single, global ability. According to Oller intelligence is mainly based on language proficiency expressed by both language and non-language abilities. But later on this view of a general or g-factor was abandoned, even by Oller.

The second phase is mainly connected with the communicative approach by Bachman and Palmer (1982) and Bachman (1990) based on Canale and Swain (1980). The Bachman & Palmer model of language ability views language proficiency as multicomponential, that is, consisting of a number of interrelated specific abilities as well as a general ability or set of general strategies or procedures. They describe two types of components: areas of knowledge which we would hypothesise to be unique to language use (as opposed to mathematical knowledge, for example) and metacognitive strategies that are probably general to mental activity (see also Celce-Murcia, Dörnyei & Thurrell, 1995; North, 1997).

Multilingual proficiency

What we will concentrate on here are specific features of multilingual proficiency in contrast to monolingual competence, which has mainly been focused on by traditional research. Researchers have admittedly applied their ideas of language competence to both FLA and SLA contexts, but have rarely referred to the differences between competence in a monolingual and competence(s) in a multilingual speaker (see Cook and Grosjean above).

In more recent discussion we have observed a multiplication of language subcomponents or skills claiming to be part of the competence originally specified by Chomsky. The number of components to be included in language competence is necessarily still an open question, particularly if we include the discussion concerning the availability of language strategies. As already mentioned before, Cummins (1979) suggests that language competence consists of the two separate skill types BICS and CALP, later labelled conversational and academic language proficiency (Cummins, 1991a: 71) (see also Chapter 2).

Whatever we might think of the current terminological controversies, one thing that will certainly have become obvious is that the original Chomskyan distinction between competence and performance – as useful and correct as it might have been at the time – is woefully inadequate in covering the plethora of components involved in the knowledge and use of a language. Based on the distinction between knowing how and knowing as suggested by Ryle (1973 [1948]) linguists have begun to differentiate between declarative and procedural language knowledge (see, e.g. Multhaup, 1997).

Although the ACT model (see Chapter 4) suggests that such a clear distinction is necessarily problematic in itself, we must assume that 'knowing a language' includes the knowledge of a language and the knowledge of how to use the language. In our view it is the latter component which is of particular significance in multilingual proficiency and/or knowledge. In an attempt at a preliminary terminological clarification we would like to suggest that competence be restricted to the field encompassed by the knowledge of a language, whilst the term proficiency – primarily derived from SLA contexts – should be reserved for the consistent outcome of the speaker's knowledge of how to use a language and the knowledge of a language. This obviously means that we cannot be assumed to use the term 'competence' in exactly the same way as Chomsky does.

L2 proficiency which refers to a learner's skill in using L2 can be contrasted with the term competence. Whilst competence refers to the knowledge of L2 a learner has internalised, proficiency refers to the learner's ability to use this knowledge for different tasks (see Ellis, 1994: 720). As the use of knowledge presupposes the presence of knowledge, proficiency is best seen as a compound form of knowledge as suggested above. Although it is not clear whether Ellis intended it to be understood in this way, the introduction of the concept of proficiency seems to us to be a necessary precondition of a meaningful discussion of language ability. This terminological clarification is, however, not intended to gloss over the essentially complex nature of the proficiency construct. This problem was, for instance, also identified by Gregg (1989: 24 quoted by Ellis, 1994: 437) who states that:

> in comparison with an attempt to construct a theory of acquisition in the domain of grammar, any attempt to construct a theory of acquisition in the domain of pragmatics or communication is going to be handicapped by the lack of well-articulated formal characterisation of the domain. On the basis of this definition we will now, nevertheless,

look at multilingual proficiency, which, as it involves involving more than one language, is necessarily more complex.

Within DMM multilingual proficiency is by necessity a derived quantity. As will be pointed out in more detail in Chapter 8 this aspect distinguishes the DMM model from other wholistic approaches, which see multilingualism as a unique phenomenon not derivable from monolingual concepts (see, e.g. Cook, 1993a; Kecskes & Papp, 2000). Whilst DMM claims that multilingual proficiency is not reducible to monolingual competence, it does see multilingual proficiency as derivable from individual language competence. Otherwise multilingual competence would probably have to be taken to derive from an innate multilingual competence ability in analogy to the language acquisition device. On the other hand we must note that proficiency is also a derived quantity in a second sense in so far as it is necessarily a hypothetical construct deduced from actual performance measured.

In DMM we start by distinguishing the following components: competence is traditionally identified as the tacit knowledge of a specific language system, which can be taken to include all the relevant syntactic and textual knowledge required; the necessarily idealised construct of multilingual proficiency describes the assumed level of acquired knowledge in the language systems of the multilingual speaker or the application of knowledge of two or more language systems. According to the above definition, this language knowledge must be taken to include the necessary procedural and declarative knowledge of more than one language system.

The hypothetical construct of multilingual proficiency is necessarily seen in terms of the language produced by speakers evaluated against an idealised endpoint, the well-educated native speaker, but this must not be understood as a return to the double monolingualism hypothesis outlined above. Underlying competence is obviously measured indirectly (see Verhoeven, 1992), whilst performance describes the actual utterances (slips, errors, etc.) of a speaker that can be influenced by stress factors, for example.

In the case of multilingual proficiency the issue is, however, further complicated by the fact that the individual language systems – or if one prefers the tacit knowledge of the respective systems – interact to produce the systematic deviation of LS_1 or LS_2, not reducible to (unsystematic) performance deviation as defined above. We must therefore assume that multilingual proficiency observes its own unique principles presented by factors unique to multilingualism.

It is quite obvious that most of the research in this field is yet to be done.

On the other hand it is clear that a wholistic approach to multilingualism requires the introduction of at least one mediating component between competence as implicit knowledge of a language and performance observed. This mediating component is referred to as multilingual proficiency. The fact that this proficiency is necessarily multicomponential does not detract from the fact that the modelling of multilingual systems requires a dynamic, pragmatically oriented component depending on the interaction between the underlying language systems and the demands of the communicative environment on the one hand and including the observed systematic variation on the other. In contrast to an analytic approach to multilingualism which presents the language systems individually, DMM provides a model of multilingual proficiency which is based on a wholistic, synthetic approach to multilingualism: the language systems involved are seen as a unity, whilst being able to specify the functions of the respective components within this model.

Current Issues in Multilingualism Research

Before we turn to the question of how to develop a realistic model of multilingualism, we will try to outline the main insights gained into the nature of multilingualism as a phenomenon in its own right, rather than a deviation from the monolingual norm or a threat to the general validity of UG.

From fractionalism to wholism

Current research on bilingualism has been greatly influenced by Grosjean's bilingual or wholistic view of bilingualism which focuses on the bilingual as a competent but specific speaker-hearer (1982, 1985). Nevertheless, many researchers still draw upon a monolingual norm assumption, interpreting bilingualism or multilingualism as a kind of double or multiple monolingualism and thus take on a perspective that would not allow a connection between the fields of research of SLA and multi-/bilingualism.

To regard SLA and bilingualism as related issues and consequently to combine the areas of SLA and bilingualism into one field of interest could be seen as fairly recent though increasing development in research (see, e.g. Harley *et al.*, 1990a; Reynolds, 1991a; Baker, 1996). This view also indicates that formal and informal SLA can lead to bilingualism. Originally two strands of research developed: sociolinguistically oriented research on bilingualism on the one hand and psycholinguistically oriented resarch on SLA on the other, the latter deriving from pedagogy and the possible (and necessary) stimulating mutual influence of the two research areas had widely been neglected.

As Grosjean (1985) mentions, the monolingual norm assumption has had an enormous impact on our concept of bilingualism and has also been (and still is) accepted by most bilinguals who criticise their own language competence or do not refer to themselves as bilingual. Only if a person is fully fluent in both languages may one call him or her a real bilingual. All other people in the world who use more than one language in their everyday life have been considered dominant bilinguals, semilinguals, etc. So researchers interested in the linguistic phenomenon of bilingualism for a long time based their studies only on ambilinguals (which are hard to find, if at all). With regard to language testing it has to be stated that tests used with bilinguals are most of the time not prepared with regard to the communicative needs or the sociocultural functions of the languages involved but according to monolingual norms, as was the case in the tests of the language skills of bilingual children in a minority language context (see Chapter 2).

It seems that the so-called negative effects of bilingualism have always been more interesting to most scholars from various backgrounds, be it educational or linguistic, than the numerous positive consequences of bilingualism. But nowadays we know that contact between the bilingual's two languages is not necessarily rare and that instances of codeswitching and of other forms of contact between the languages are characteristic of a bilingual's speech. Bilinguals normally mix their languages in their everyday speech even when they are fully fluent in both and are also able to use them separately (see Milroy & Muysken, 1995).

Grosjean's attempt to present the bilingual speaker in a bilingual or wholistic approach has influenced the scientific debate on multilingualism. To him the bilingual (or multilingual) speaker can be compared to a high hurdler who combines the two types of competences, jumping and sprinting, in one person, although s/he is neither a sprinter nor a high jumper. The bilingual is a human communicator who has developed a commmunicative competence sufficient for everyday life. The bilingual is not the sum of two complete or incomplete monolinguals; s/he rather has a specific linguistic configuration characterised by the constant interaction and co-existence of the two languages involved. The bilingual's competence thus cannot be evaluated with the traditional language testing methods but has to be studied through the bilingual's psycholinguistic system which may also be characterised by mixed speech (see Grosjean 1985: 471).

With regard to Grosjean, Cook (e.g. 1993a, b) suggests referring to the knowledge of L1 and L2 as to multicompetence and thereby criticises the approach taken in UG research, which starts from the monolingual speaker as the norm. As already argued in Chapter 4, Cook (1993b: 3–4) sees several good reasons why a proper account of second language learning will have to treat multicompetence in its own terms, not in terms related to double monolingualism. That is, in contrast to monolinguals, bilinguals or multilinguals have a different knowledge of their first language, their second language, a different kind of language awareness and language processing system.

Linguistic, cognitive and sociocultural consequences of multilingualism

Crosslinguistic interaction and other drawbacks in multilinguals

From the outset it is to be noted that being multilingual does have a number of drawbacks which have little to do with the disadvantages suggested by early researchers (see, e.g. Hakuta, 1986).

> It is, of course, an advantage for a child to be familiar with two languages but without doubt the advantage may be, and generally is, purchased too dear. First of all the child in question hardly learns either of the two languages as perfectly as he would have done if he had limited himself to one. It may seem on the surface, as if he talked just like a native, but he does not really command the fine points of language [...]. Secondly, the brain effort required to master the two languages instead of one certainly diminishes the child's power of learning other things which might and ought to be learnt. (Jespersen, 1922: 48)

Multilinguals (particularly these are taken to include second language learners) appear to find maintaining and managing more than one language quite a challenge. Thus Hamers and Blanc (1989: 107) report on interesting research on the stress related language performance of multilinguals, indirectly supporting the idea that language management forms an essential part of multilingual performance. This observation is confirmed by the evidence for deficit characterised by Cook (1993a: 111) as follows: 'L2 learners perform less well in the L2 than equivalent native speakers across the board. Anything involving processes of speech production, comprehension, memory, information storage, and so on, shows a deficit compared to natives.' If we add to this the observation of the multilingual underachievement already commented on, we will have to admit to the existence of a general phenomenon requiring explanation.

As was argued in Chapter 3, the attribution of multilingual underachievement to interference may at first sight seem plausible. The interference effects are, however, obviously not simply negative but multidimensional even if we accept the provisional distinction between (negative) transfer and crosslinguistic influence. Whilst there is in some instances evidence of one language system interfering with, and possibly its mere existence restricting the accessibility of, the other language system, the interpretation of crosslinguistic effects as primarily negative in terms of reducing the respective language achievements of the multilingal speaker represents not only a very one-sided view of the effects to be expected but also constitutes a misunderstanding of the nature of the multilingual's language system. As we have tried to show in the chapter on transfer, we regard the contact between the languages in a multilingual speaker as characterised by a range of transfer phenomena which in Chapter 3 we subsumed under crosslinguistic interaction (CLIN) to suggest an extension of the concept of crosslinguistic influence (CLI) as the commonly used concept in language learning studies.

Benefits from multilingualism

In contrast to early research which showed bilinguals to be greatly disadvantaged in comparison to monolinguals, many studies in the last twenty-five years or so have shown that multilingualism can result in advantages. Scholars involved in the study of bilingualism have moved from an enthusiastic attitude towards bilingualism and the cognitive advantages gained from it to a more diversified but nevertheless optimistic view of bilingualism (see Hakuta, 1986; Reynolds, 1991b) largely influenced by the positive results of the Peal & Lambert study which revealed the paradox of transfer, a transfer phenomenon affecting both the language and the cognitive style of the subjects. In the meantime, it has become widely known that under specific conditions multilinguals can have tremendous advantages over matched monolinguals, not only in terms of language competences, but also in terms of cognitive and social development.

> The assumption behind all four mechanisms is that the primary effect of bilingualism is on language-learning and processing strategies, and that it is through this channel that bilingualism may affect general thought processes. These mechanisms all involve resolving interference at the structural level of language. They are as follows: (1) *language analysis*; (2) *sensitivity to feedback cues* indicating correctness or incorrectness of present language orientation; (3) *maximisation of structural*

differences between languages; (4) *neutralisation of structure within a language.* (Ben-Zeev, 1977: 31)

Within our concept of multilingual proficiency we would like to stress the importance of multilingual skills as developed in the multilingual speaker whose procedural knowledge shows several characteristics which clearly distinguish her/him from the monolingual speaker. The acquisition of more than two language systems leads to the development of new skills such as metacognitive strategies due to the learner's experience of learning how to learn a language and an enhanced level of metalinguistic awareness.

As already pointed out earlier on, a number of studies have not only reported a bilingual superiority on measures derived from various cognitive skills but in some cases, positive crosslinguistic relationships (i.e. also from L2 to L1) have been found for pragmatic or conversationally-oriented language abilities in addition to literacy-related abilities as suggested by Cummins (e.g. 1991b) and Kecskes & Papp (2000) (see Chapter 3). Considerable evidence shows that the development of proficiency in two languages can result in greater levels of metalinguistic awareness and the facilitation of additional language acquisition by exploring the cognitive and linguistic mechanisms underlying these processes of transfer and enhancement which are closely linked to linguistic distance and psychotypology (see, e.g. Kecskes & Papp, 2000: 87–105). Many studies have strongly indicated that bilinguals show definite advantages on measures of metalinguistic awareness, cognitive flexibility and creativity (see, e.g. Baker, 1996: 128–161) and therefore differ in thinking styles from their monolingual counterparts. Additional insight into the relationship between metalinguistic behaviour and bilingualism can also be gained by studies on translation skills of bilinguals (see, e.g. Malakoff, 1992).

Metalinguistic awareness

The ability to focus attention on language as an object in itself or to think abstractly about language and, consequently, to play with language is one of the features typical of a multilingual's cognitive style in contrast to most monolinguals' (see Bialystok, 1991b: 114). The research on a multilingual's most characteristic cognitive ability, that is linguistic objectivation, must, however, be seen as rather heterogeneous.

A close look at the literature on metalinguistic behaviour makes clear that the terminology used in this growing area of research on multilingualism is actually rather confusing (see Jessner, 1995: 177). We are confronted with metalinguistic awareness, metalinguistic skills, metalinguistic abilities, metalinguistic tasks, all terms which are not used systematically (see Bialystok, 1991b: 114). For our purposes we would like to follow Gombert (1992: 13) who defines metalinguistic activities as 'a subfield of metacognition concerned with language and its use – in other words comprising: (1) activities of reflection on language and its use and (2) subjects' ability intentionally to monitor and plan their own methods of linguistic processing (in both comprehension and production).'

In her approach to bilingual language processing, Bialystok (1991b; 1992) focuses on analysis and control as the metalinguistic dimensions of language proficiency. These information-processing components are the processes defining performance across tasks which determine entry into the metalinguistic domain. Bialystok reports on evidence from different studies of bilingual children who turned out to be able to solve problems in three language domains better than their monolingual peers because of different levels of mastery of analysis and control processes based on their bilingual experience. Bialystok concludes that there are no universal advantages, but that the processing systems developed to serve two linguistic systems are necessarily different from the processing systems that operate in the service of only one. Thus, bilinguals who have attained high levels of proficiency in both languages are viewed as being advantaged on tasks which require more analysed linguistic knowledge.

Most studies focus on the (bilingual) child and the onset of metalinguistic awareness (see, e.g. Tunmer & Pratt & Herriman, 1984). Of the very few researchers who have discussed the development of metalinguistic abilities Karmiloff-Smith (1992) and Gombert (1992) have to be mentioned in the context of monolingual competence in first language acquisition. In his international research project Titone (1994) has worked on the development of metalinguistic knowledge in multilinguals. He distinguishes between language awareness in young children and metalinguistic consciousness, which in his opinion develops after the age of twelve years and in children growing up in a bilingual family. However, as pointed out by Hoffmann (1991: 80–93) it is considered difficult to find out about the onset of metalinguistic awareness in bilingual children since some of them do not talk explicitly about their language systems but would show their metalinguistic awareness in their linguistic behaviour.

Metalinguistic awareness in multilinguals should be regarded as closely linked to the idea of monitoring. In SLA research the monitor has mainly been related to Krashen's idea of a monitor in the language learner which is defined as 'that part of the learner's system that consciously inspects and, from time to time, alters the form of the learner's production.' (Dulay &

Burt & Krashen, 1982: 279) In our model the multilingual speaker's system is characterised by an enhanced multilingual monitor, that is, the monitor is used by the multilingual speaker to watch and correct her or his language(s) in a multilingual context which increases the use of such a monitor. Indeed, there is some evidence to suggest that the language monitor experiences a significant development in multilingual systems. With the increase in the number of languages involved the functions of the monitor expand.

Such an extended language monitor can be conceived of as having the following significant functions:

(1) fulfilling the common monitoring functions (i.e. reducing the number of performance errors, correcting misunderstandings, developing and applying conversational strategies);
(2) drawing on common resources in the use of more than one language system and
(3) keeping the systems apart by checking for possible transfer phenomena and eliminating them and thereby fulfilling a separator and cross-checker function.

The multilingual individual habitually transfers elements from one language to another and forms rules according to commonalities and differences in her or his language systems (see, e.g. Jessner, 1999).

Cognitive flexibility

In addition to the studies of metalinguistic awareness, many investigations have indicated that young bilingual people, compared to monolingual controls, show definite advantages in cognitive flexibility, creativity, divergent thought or problem solving. Multilinguals and monolinguals have turned out to differ in thinking styles. Bilingual children proved to be divergent thinkers, that is, they are more creative, imaginative, flexible and unrestrained in their thinking. In tests of creative thinking, bilinguals have outscored their monolingual counterparts with regard to fluency, flexibility, originality and elaboration. In a study on the cognitive development of Italian-English bilinguals and Italian monolinguals, Ricciardelli (1992), for instance, found that bilinguals who had attained a high level of proficiency in both languages performed significantly better on creativity, metalinguistic awareness and reading thus supporting Cummins' threshold hypothesis (see Chapter 2).

Metapragmatic and sociocultural awareness

The use of two or more languages not only influences the linguistic and cognitive skills of multilinguals, but also their social skills. In contrast to van Kleeck (1982), who emphasises the social use of language referring to metapragmatic activity, we do not wish to exclude metapragmatic behaviour from metalinguistics but stress the importance of pragmatic competence or communicative sensitivity as components of sociocultural knowledge which seem to be more developed in multilingual than in monolingual speakers. Which clues, for example, does a multilingual person need in order to know which language to choose in which situation or how to follow socioculturally determined discourse patterns in a conversation with other multilinguals?

The field of metalinguistics concerns all levels of language, but in a language contact situation the pragmatic aspect of language seems to be of utmost importance. In a model of multilingualism special emphasis has to be put on crosscultural pragmatics or interlanguage pragmatics (see, e.g. Kasper & Blum-Kulka, 1993). Differing situations call for different linguistic skills, that is, the speaker can use routines and patterns in some situations but is forced to use complex linguistic structures in others. Several studies report that bilinguals are more sensitive than monolinguals in interpersonal communication. In their classic study investigating children from five to eight on their performance in the explanation of a dice game, Genesee, Tucker and Lambert (1975) could prove the superiority of bilingual subjects with regard to communicative sensitivity. The bilingual children's explanations of the game turned out to be more appropriate to the listeners' needs than the explanations given by the monolinguals. Communicative sensitivity towards the needs of the communication partners appears to be related to the concept of interactional competence, which refers to the ability of a person to perform and interpret communicative actions, for example, thanking and greeting, in interactional situations, with regard to the sociocultural and sociopsychological norms of the community involved (see Oksaar, 1990).

All in all, trends suggesting multilingual superiority in aspects of metalinguistic development and sociocultural awareness have been so impressive in studies carried out during the past two to three decades that we must assume that metalinguistic awareness is one of the factors in which multilinguals outperform their matched monolingual counterparts.

Studies in third language acquisition and trilingualism

This field of research represents a rather young discipline within linguistics which has, however, been gaining more and more interest over the last years (see Clyne, 1997; Cenoz & Genesee, 1998; Hufeisen & Lindemann, 1998; Cenoz & Jessner, 2000; Dentler & Hufeisen &

Lindemann, 2000). Although the number of studies on third or fourth (or more) language acquisition is still very limited, this research area has already established itself as a field of its own by emphasising the differences between TLA and SLA as well as pointing out that other aspects of learning a third language have to be seen as similar to SLA.

Since research on TLA and trilingualism is based on studies on bilingualism and SLA psycholinguistic models of the two disciplines have been applied to explore the trilingual system. In quite a number of studies on multilingual processing, where the majority is concerned with the trilingual lexicon (see, e.g. several contributions in Cenoz & Hufeisen & Jessner, 2001), researchers mainly draw on Green's model for the control of speech in bilinguals (1986) and de Bot's bilingual production model based on Level's model of speech processing (1992) as sometimes complementary research tools. According to Green the two languages in a bilingual have different levels of activation and when a language is selected and controls the output, the highest level of activation occurs. He differentiates between selected, active and dormant languages. By drawing on Green, de Bot applies Levelt's ideas on human speaking, where utterances are successively produced in a conceptualiser, a formulator and an articulator, to the bilingual speaker (see also Poulisse, 1997). In a different approach Paivio (e.g. 1986) suggests that the bilingual has two separate verbal language systems and a separate non-verbal image system which is independent from the language systems in a bilingual and serves as a shared conceptual system for the two languages. There are also interconnections comprising association and translation systems to be found between the two lexicons of a bilingual speaker.

Grosjean's work on the language mode continuum (1992) has made clear that the control variables commonly used in bilingualism research have to be extended by language mode in order to provide better insight into the variability of speech situations. He describes language mode as the state of activation of the bilingual's languages and language mechanisms at a given point in time and in a more recent publication also applies his ideas to trilingual speakers by pointing out that a trilingual speaker can certainly be imagined in a monolingual, a bilingual or a trilingual mode with various levels of activation (2001).

Whereas in second language learners two systems can influence each other the contact between three language systems in a multilingual speaker can develop more forms, that is, apart from the bidirectional relationship between L1 and L2, L3 can influence L1 and vice versa and also L2 and L3 can influence each other. Thus crosslinguistic aspects of TLA and trilingualism (see, e.g. Clyne, 1997; Cenoz & Hufeisen & Jessner, 2001), form an important part of the topics discussed in studies of TLA and trilingualism. Williams and Hammarberg (1998) present several criteria which they consider influential in the relationship between the languages in L3 production and acquisition: typological similarity, cultural similarity, proficiency, recency and the status of L2. The role of L2 in TLA has turned out to be of greater importance than originally suggested. As an extension of interlanguage theory as applied to SLA (see Chapters 2, 3) de Angelis and Selinker (2001) have introduced interlanguage transfer as the influence from a non-native language on another non-native language into the multilingualism discussion. The study of language attrition in trilingual speakers (Cohen, 1989) and re-emergence of (two) languages in a multilingual (Faingold, 1999) will hopefully serve as stimulating examples for more work in these areas in the future.

Another important aspect is the rather small albeit growing field of child trilingualism (see Hoffmann, 1985; Arnberg, 1987; Hoffmann and Widdicombe, 1998; Barron-Hauwaert, 2000; Gatto, 2000) which is seen as linked to educational aspects such as trilingual education in primary school (see, e.g. Cenoz & Lindsay, 1994; Ytsma, 2000). The question whether the acquisition of a further language is facilitated in bilingual speakers can be seen as one of the key aspects of the research conducted in the field.

In some studies bilingualism has already been stated to be beneficial in the learning process of the third language. In a Basque study Cenoz and Valencia (1994) were able to show that monolingual Spanish students were outperformed by their bilingual Basque colleagues in the acquisition of English as their third language (see also Lasagabaster, 1997). Similarly, Ringbom (1987), one of the first scholars to show interest in TLA, reports of the advantages of Swedish speaking Finns over monolingual Finnish students when acquiring English in Finland. Ringbom concludes that among other factors like language typology or linguistic experience resulting from bilingualism which play an important role in this particular language learning context, the high degree of automaticity as found in expert learners is influential.

Information processing strategies and techniques in expert language learners differ from the controlled processes governing language learning in novices, that is in more experienced learners routinised complex skills become automatic after the earlier use of controlled processes (see McLaughlin & Nayak, 1989: 7). In an experiment comparing the information-processing skills of multilinguals and monolinguals the multilingual subjects outperformed the monolinguals on the implicit-learning task because of better automated letter- and pattern-recognition skills (see Nation & McLaughlin, 1986).

In general, interest in research on strategy use in (second) language learning is also a growing field as a number of recent publications have shown (e.g. Oxford, 1990a; Bialystok, 1990; O'Malley & Chamot, 1990; Kasper & Kellerman, 1999; McDonough; 1999). The importance of research on language learning strategies becomes even more obvious in a context involving more than two languages since the change in quality of the strategies in learners with experience in prior language learning leads to the catalytic or speeding up effects as known from multilingual learning. Apart from earlier publications on the good and exceptional language learner (see, e.g. Naiman & Fröhlich & Stern & Todesco, 1996 [1978]; Gillette, 1987) we have very little evidence on the nature of learning strategies applied by multilinguals (see Hufeisen, 2000a; Mißler, 2000) so that this issue certainly needs further investigation to be of use in a world of growing multilingualism.

Classification and terminology in multilingualism studies

It is important to note that the study of more than two language systems has made clear that the traditional terminology in SLA studies cannot be applied without questioning to studies on TLA and trilingualism, that is the common use of the symbols L1 and L2 in connection with FLA and SLA have to be reconsidered in a research context going beyond two languages. This is not to say that this issue has been ignored by scholars involved in SLA research but to emphasise the importance of such a distinction in a multilingual context. When we use L1 and L2 to describe the relationship of the two languages involved in a bilingual system L1 is usually interpreted not just as the first language learnt but also as the dominant language. This distinction can lead to confusion when L1 is not the dominant language any longer, that is, when due to changes in communicative needs L2 has become the dominant language of the speaker. In such a case it has been suggested that the term 'primary language acquisition' (henceforth PLA) be used instead of FLA when referring to the language learning process of the first language learnt (in a monolingual environment).

Already from this discussion it becomes clear that the command and/or use of more than two languages can result in even greater or more changes in the individual psycholinguistic system, a fact that has to be acknowledged in multilingual studies. Whenever we describe the languages in a multilingual speaker we have to make sure that a distinction between the onset of learning the languages and the dominance of the languages involved has to be reflected in the terminology used.

Furthermore, as already pointed out by Hufeisen (2000b: 211–212), it is sometimes difficult for a multilingual speaker to describe her/his language competences by the use of chronological terminology since the competences usually change over time and in addition skills within the languages can vary. Since a basis for academic discussions is needed this is an important issue to be readdressed in more detail in the future. As can be seen in the next chapters, to indicate our awareness of the delicate issue of classifications used in multilingualism studies we have decided to use PLA in our model of multilingualism.

Multilingual Variation

Obviously the scope of language variation is an aspect in which multilinguals differ greatly from monolingual speakers. Having command of more than one language has the obvious consequence of language choice, that is, the speaker can choose the language of communication according to the requirements of the respective situation. As can be gathered from the discussion on CS and, for instance, Grosjean's contributions on the language mode continuum as presented above, multilinguals make ample use of that choice. Thus multilinguals exhibit a greater linguistic variety measured in terms of monolingual performance.

The most frequent view is the interpretation of variation as a purely sociolinguistic phenomenon, which we will term interpersonal variation. Interpersonal variability refers to the fact that different speakers of the same language show considerable variations in terms of the phonetic, syntactic and lexical systems of a language. More recently attention has turned towards intrapersonal variation, that is variation as a psycholinguistic phenomenon. Yet even here we can distinguish between at least two types of variation, the first being variation in register depending on the respective sociolinguistic situation. We can assume that variation in register basically follows the same principles as language mixing outlined in Chapter 3. That is, intrapersonal variation – like CS – is first and foremost understood as a performance phenomenon, and this not only by UG researchers (see Chapter 4) but also, for example, by Edwards in his book on multilingualism (1994) whose chapter on variation within a language mainly covers situational variation with passing reference to accommodation and diversification.

The second more difficult aspect of variation is that of uneven performance, the fact that even native speakers systematically fail to achieve the level of performance predicted by their assumed language competence. This problem is exemplified by the fact that native speaker grammaticality judgements are not as consistent as we would predict on the basis of (UG-oriented) linguistic theory. This form of intrapersonal variation is obviously even more pronounced in L2 learners (see Ellis, 1994: 119–159).

Variation has turned out to play an important role in various research designs and approaches to SLA. Several theories have been developed to deal with variability in language competence on the one hand, and with systematicity on the other. As mentioned above, linguists following the Chomskyan paradigm have adopted a homogeneous competence model whereby variation is seen as a feature of performance rather than of the learner's underlying knowledge system. Homogeneous linguistic competence is focused on as the main goal of inquiry and consequently variability is ignored by this paradigm (see Salkie, 1990: 62).

In principle, sociolinguistic and psycholinguistic approaches do not ignore variability: psycholinguistic mechanisms are used to explain how situational factors result in variation, often complementary with sociolinguistic factors. According to Ellis (1994: 120) these approaches are complementary and a full account of variability requires both. To perceive the speaker of a language as having a finite state competence, as suggested by UG linguistics, makes it quite impossible to understand the social dynamics in a multilingual community resulting in changes of competence in the individual, which often result in language attrition or loss. This linguistic phenomenon, which will be explained in more detail in Chapter 6, has been shown in many sociolinguistic studies and in some psycholinguistic ones published fairly recently which mainly concentrate on the contact between two languages (see, e.g. Weltens & de Bot & van Els, 1986; Seliger & Vago, 1991; Hyltenstam & Viberg, 1993). The number of studies in this area focusing on three or more languages is still very limited (e.g. Cohen, 1989; Faingold, 1999).

Diachronic changes on an individual level seem to be worth investigating, especially when we consider the question of the temporary, as opposed to permanent nature of positive effects of multilingualism on the cognitive development of multilinguals (Baker, 1996: 142). As studies of multilingualism and its impact on the cognitive system have mainly been concerned with the child, further research is needed to find out whether the positive impact of early multilingualism on cognitive growth will also show in adult multiltilinguals or whether monolingual controls will catch up with their multilingual counterparts.

The central empirical focus changes from an attempt to test absolute models of competence to an attempt to predict and test which components of competence will become differentiated from each other for particular groups of language learners in specific acquisition and learning contexts (see, e.g. Jisa, 1999). These predictions for empirical testing would draw on both abstract models of competence and theories regarding the developmental effects of individual, social, and instructional variables on second language learning. Developmental changes in the individual's language proficiency as the source of variability are focused on in a dynamic view of the language acquisition process.

Sociolinguistic variation

The fact of language variation as such has been known since the birth of linguistics. It has always been assumed that language can or does change over time. Regional variation is also a well-known fact of language communities. Explanations of language variation would therefore rely on arguments based on diversification and changes in the language habits of respective generations of speakers and not in terms of changes in the language or competence of a specific language speaker.

The traditional approach to variation in language, if we ignore the aspect of language change, is a sociolinguistic one. Language change or variation has therefore been seen primarily as a sociolinguistic phenomenon. Consequently language variation has automatically been interpreted as inter-speaker variation, that is in terms of regional variation or sociolect. And even then it is assumed that sociolinguistic variation does follow very specific rules: '[V]ariation is not random but strictly controlled, often by extra-linguistic factors, and the specification of these factors may help us account for the change.' (McMahon, 1994: 226).

Since Labov's much-cited investigations of language change within a language community (e.g. 1969) linguists have accepted that individual language patterns can change considerably, as long as they observe certain limits (of the language system). It is also accepted that the factors determining language change are not merely internal linguistic ones but rather of a social and psychological nature. In contrast to the view suggested by UG language in use appears to be subject to constant variation frequently affecting whole communities of speakers. The first dynamic sociolinguistic models were suggested by Bailey (1973) and Bickerton (1975). What is known as wave theory suggests that language changes spread through a speaker community leading to synchronic variation in individual speakers, to be interpreted as a transition from one rule system to another. The speaker thus adapts to a changing environment by modifying her/his rule system to accommodate to the altered environment.

Psycholinguistic variation

Let us first of all call to mind the conventional UG approach to variation as already discussed in Chapter 4. At first sight UG's treatment of variablity in the speaker seems to be an obvious one. The speaker commands a homogeneous competence (see, e.g. Tarone, 1983) and the problem of variablity (as originally suggested by Chomsky) is relegated to the sphere of performance, which means that any variation in speaker competence is impossible.

As already mentioned in Chapter 4, evidence of psycholinguistic variability on an individual level (i.e. systematic speaker and learner variation) appears to have increased and therefore even UG-influenced researchers are forced to take this challenge rather more seriously. Whereas synchronic variation is compatible with a homogeneous competence model – all that it requires is an extension of competence to include the possiblitiy of CS between a number of registers, sociolects or dialects – diachronic variation has not only to include the possibility of positive and negative growth within the speaker's language systems, that is the grammatical, lexical, phonetic system, etc., but also has to be linked with individual factors of a psychosocial nature in language learners.

The complexity of multilingual systems is partly due to the numbers of individual factors, such as attitude and motivation, anxiety, to name but a few, which have been identified as crucial in the (second) language learning process (see, e.g. Skehan, 1989; Kuhs, 1989; Gardner & MacIntyre, 1992) and which are also subject to change on the individual level. One significant psycholinguistic variable in the development of multilingual systems is language aptitude.

On the role of language aptitude

Variation in the language learning progress has frequently been attributed to differing language aptitude which has been discussed extensively and controversially in a language learning context (see Carroll & Sapon, 1959; Diller, 1981; Vollmer, 1982; Skehan, 1989; Parry & Stansfield, 1990). Whereas most studies of individual factors in language learning concentrate on a single factor and its importance for the language learning process Gardner, Tremblay and Masgoret (1997) have included language aptitude in their model of SLA and describe its relationships to language anxiety, attitudes and motivation, field dependence/independence, language learning strategies and self-confidence. The results of the study indicate substantial links, but the role that language aptitude or (multi)language aptitude can play in a the study of multilingualism remains, nevertheless, to be clearly stated. In the recent discussion of language aptitude learning styles and strategies have been treated as possible components, or at least as potential correlates (see Oxford, 1990b: 67).

A useful definition of language aptitude is provided by Carroll (1981: 84):

> Aptitude as a concept corresponds to the notion that in approaching a particular learning task or programme the individual may be thought of as possessing some current state of capability for learning that task – if the individual is motivated, and has the opportunity to do so. That capability is presumed to depend on more or less enduring characteristics of the individual.

In their original research Carroll and Sapon (1959) described aptitude in the following terms: (1) sound coding ability: the ability to identify and remember new sounds in a foreign language, (2) grammatical coding ability: the ability to identify the grammatical functions of different parts of sentences, (3) inductive learning ability: the ability to work out meanings without explanation in a new language; and finally (4) memorisation: the ability to remember words, rules, etc. in a new language.

Language aptitude has, however, been largely ignored in more recent linguistics, possibly because psycholinguistic findings have made the existence of a general psycholinguistic entity such as language aptitude appear doubtful. It has become commonplace to interpret enduring characteristics as meaning innate properties, which was the original interpretation offered by Carroll, claiming that aptitude cannot be trained. Whether language aptitude changes with language experience in relation to age (see Eisenstein, 1980; Mc Laughlin, 1990; Harley & Hart, 1997) or can be seen as a stable component is part of an ongoing discussion but there seem to be some factors on which research tends to agree.

First, aptitude is considered a separate component and is not identical with intelligence (as originally suggested by Neufeld, 1979). Aptitude is also generally considered a valid predictor of the rate of learning of a language (see Skehan, 1998: 185–235), and it is in this respect that language aptitude is of interest in multilingual systems.

Should this psycholinguistic category be taken to exist, we can possibly conduct tests to ascertain whether multilingual speakers perform better in standard language aptitude tests. Yet even if the existence of a general category language aptitude should not be confirmed, we can rephrase the question as to the positive influence bilingualism can be taken to have on TLA. To determine the relationship of the concepts of metalinguistic awareness and (multi)language aptitude and discuss this issue with regard to developmental change in the learner could present a most challenging enterprise for future research on multilingualism. The evidence so far suggests that aptitude is a property in its own right and does function as a valid predictor of the rate of language learning. It is also quite obvious that the property of language aptitude varies in individual speakers and is generally taken as given in language learning. As noted before it represents only one, albeit an important one, out of many factors exerting influence on the process of language learning.

Arguments for a Dynamic View

The dynamics of the processes involved in individual progression and regression and the complex interdependences between the factors involved in the language acquisition process are focused on in a dynamic view of language acquisition and multilingualism. In DMM we stress the fact that language changes in time on an individual level.

As Hyltenstam and Viberg (1993: 3) point out, language can be seen as inherently dynamic, something that exhibits change and flux, and is characterised by motion. In contrast to fossilised languages, living languages are in a continuous motion. They adapt to the social contexts in which they are used and they move with time, changing chronologically. A specific instance of chronological or diachronic change occurs on the individual level.

Language change in the individual results from adjusting one's language system(s) to one's communicative needs. If, like Grosjean (1985: 473), you look at the bilingual as an integrated whole, you can watch how changes in the language environment, and therefore in language needs, affect her/his linguistic competence in the one or the other language, not in her/his linguistic competence in general. Speakers may move from monolingualism to bilingualism, from bilingualism to trilingualism, that is different systems (LS_1, LS_2, LS_3, etc.) are transitionally commanded by the same individual. According to the communicative needs, the native speaker has transitional command of different language systems over a period of time, resulting, for instance, in monolingualism, bilingualism, trilingualism, etc.

We have developed a psycholinguistic working model of multilingual proficiency on the basis of monolingual growth models. Whereas Chomskyan linguistics describes language competence as an invariable state, we argue that language competence as such should be seen as neither an absolute nor invariable state, but as an attainable goal, both for native (!) and non-native speakers. According to Nemser's theory of approximative systems, linguistic systems can be seen as transient stages, even if the target language is never attained. This means that at least some second language learners never actually learn their target language, but that these learners at some stage stop developing their foreign language skills and settle for what they have got. We will recognise this phenomenon as something we have already encountered in a different context (see Chapters 2 and 4) and which was then referred to as fossilisation.

Furthermore, our psycholinguistic model of multilingualism is learner-oriented and tries to explain individual learner differences in language acquisition. We are interested in various factors affecting (the learner's) language performance, for instance, attitude and motivation, anxiety, language aptitude. The approach taken in our model thus focuses on the dynamic systems of language learners. This view implies that language itself is in constant flow, and so are the language systems in a multilingual, depending on the various factors involved in the language acquisition process.

According to Grosjean (1985: 473), it is extremely interesting to study the wax and wane of languages in a bilingual person, and therefore of language needs, which affect her or his linguistic competence in the one language and in the other, but not in her or his competence in general. Generally, we assume that a person can go in and out of bilingualism, can shift totally from one language to the other in the sense of acquiring another language and forgetting the first totally, but will never depart (except in transitional periods of language learning and forgetting) from a necessary level of communicative proficiency demanded by the environment (see Chapter 7). S/he will develop competence in each of her or his languages to the extent required by the environment where the competence in one language may be quite rudimentary.

DMM provides the necessary conceptual psycholinguistic framework for modelling multilingual proficiency, putting special emphasis on individual learner differences in language acquisition. It describes the language systems of a bi- or multilingual reacting differently to identical input in different situations, that is, different languages commanded by the same speaker which are viewed as separate systems (LS_1, LS_2, LS_3, etc.) exhibiting different properties. This model, taking the wholistic view of bilingualism into account, stresses the fact that an adequate description of multilingualism must comprise not only transfer phenomena including codeswitching, language mixing, language attrition, but also the positive cognitive consequences of multilingualism (e.g. enhanced metalinguistic and metacognitive abilities, divergent thinking), which become apparent if certain social and cognitive conditions are met. Multilingual proficiency is, therefore, to be considered as consisting of dynamically interacting linguistic subsystems which themselves do not necessarily represent any kind of constant but are subject to variation.

Chapter 6
A Dynamic Model of Multilingualism Developed

In the following we will outline some general aspects of modelling multilingualism within a dynamic systems concept. An introduction to some of the fundamental ideas of systems-theoretic approaches will be provided and put in relation to our research. The promises such an approach to multilingualism offers to research on language acquisition theory and MR will be explained in more detail in the next section. The process of building a dynamic model will then be discussed with a look at the goals of DMM, the introduction of variables and constants and the presentation of a portrait of a dynamic multilingual system. Gradual language loss and language maintenance will be identified as the most important essential factors of such a system. Furthermore predictions concerning multilinguals' advantages and disadvantages made in the model will be considered in this overview of the model's development. Finally working hypotheses concerning language growth and proficiency will have to be introduced before starting to develop the model in more detail in the next chapter.

The relationships between the parameters considered relevant to the multilingual system are both explicated in the text and presented in graphs or graphical representations, as to our minds it is not merely important to understand the relationship between, say, two factors but also to understand how they contribute to a dynamic development of the system in time.

Introduction to Dynamic Systems Approaches

For several decades dynamic systems have caught the conceptual interest of researchers from various fields of science. Since its inception systems theory research has grown into a vast field of knowledge ranging from cybernetics and neuroscience to mathematical models of change as found in chaos theory. The biological line of thought stems from early biologists such as Uexkuell (1973) to more recent holistic concepts presented by Waddington (1977) and many others. This development can be traced through von Bertalanffy (1968), Bateson (1972, 1979), Haken (1981) to Casti

(1991), Nørretranders (1997), Dörner (1992), Waldorp (1992), Braitenberg and Hosp (1994), Crick (1994), Basieux (1995), Gottman (1995) and Simon (1997), to name but a few, whilst ignoring the theoretical developments leading to radical constructivism.

In his book on the history of research on chaos theory Gleick (1987) describes the discoveries and scholars involved in the development of the field crossing the boundaries of scientific disciplines. The approach to reality expressed in chaos theory actually dates back to ancient history described in Gleick's (1987) as well as in Briggs and Peat's (1989) excellent introductions to the field.

In its varieties and forms the systems idea has influenced more or less every discipline. Thirty years ago there

> were mathematicians, physicists, biologists, chemists, all seeking connections between different kinds of irregularity. Physiologists found a surprising order in the chaos that develops in the human heart, the prime cause of sudden, unexplained death. Ecologists explored the rise and fall of gypsy moth populations. Economists dug out old stock price data and tried a new kind of analysis. The insights that emerged led directly into the natural world – the shapes of clouds, the paths of lightning, the microscopic intertwining of blood vessels, the galactic clustering of stars.

As a consequence by the middle of the eighties departments of systems research were installed at university level almost all over the world (see Gleick, 1987: 38; Briggs & Peat, 1989: 175) and 'chaos has become a shorthand name for a fastgrowing movement that is reshaping the fabric of the scientific establishment.' (Gleick, 1987: 3)

At the core of the theory is the understanding of the behaviour and organisation of living organisms as dynamic systems.

> A system […] is more than just a collection of variables or observables we have isolated from the rest of the world. It is a system primarily because the variables mutually interact. That is, each variable affects all the other variables contained in the system, and thus also affects itself. This is a property we may call complete connectedness and it is the default property of any system. The principal distinctive property – compared to a constant – is that it changes over time. Consequently, mutual interaction among variables implies that they influence and co-determine each other's changes over time. In this sense, *a system is, by definition, a dynamic system* and so <u>we define</u> a dynamic system as a

set of variables that mutually affect each other's changes over time. (van Geert, 1994: 50; italics inserted by the authors)

Much of the work of the researchers can be seen as an attempt to describe, model and predict the activity, including the controlling factors, of time-dependent changes in the system. In order to find out about some underlying lawfulness in the phenomena studied, the organisation and pattern of complex systems are analysed and attempts are made to discover the responsibility for stability and instability and the nature of structural reorganisation (see Robertson & Cohen & Mayer-Kress, 1993: 119). One of the advantages of dynamic systems theory is that it can provide the research community with rich metaphors in our attempt to understand behaviour and development (see Robertson & Cohen & Mayer-Kress, 1993: 148).

Most biological systems and many physical systems can be described as discontinuous, inhomogeneous and irregular, that is, turbulence, irregularity, and unpredictability are everywhere, but it has always seemed fair to assume that this was 'noise' resulting from the variable and complicated structure of the behaviour of living systems in reality. In theory this messiness could be reduced to its orderly underpinnings.

> Essentially reductionism is a watchmaker's view of nature. A watch can be disassembled into its component cogs, levers, springs, and gears. It can also be assembled from these parts. Reductionism imagines nature as equally capable of being assembled. Reductionists think of the most complex systems as made out of the atomic and subatomic equivalents of springs, cogs, and levers which have been combined by nature in countless ingenious ways. (Briggs & Peat, 1989: 21–22)

The reductionist view is clearly expressed in linear relationships as part of the theory. As Gleick (1987: 23) puts it:

> Linear relationships can be captured with a stright line on a graph. Linear relationships are easy to think about: the more the merrier. Linear equations are solvable, which makes them suitable for textbooks. Linear systems have an important modular virtue: you can take them apart and put them together again – the pieces add up.

Bates and Carnevale (1992: 9) provide the following formula for linear relationships:

> 'By definition, a relationship between two variables is linear if it can be fit by a formula of the type

$$y = ax + b$$

where y and x are variables, and a and b are constants. Any relationship that cannot be fit by a formula of this kind is, by definition, non-linear.'

According to Gleick (1987: 23): '[n]onlinear systems generally cannot be solved and cannot be added together. In fluid systems and mechanical systems, the non-linear terms tend to be the features that people want to leave out when they try to get a good, simple understanding.' This seems to support the viewpoint of structural theory viewing change and variability in real time as noise and thus leaving it unexplained (see Smith & Thelen, 1993: 163). As argued by Aslin (1993: 397) dynamic systems theory 'points to the potential importance of variability, not as error variance, but rather as a lightning rod for studies of critical points during development and as a means of creating opportunities for developmental change.'

Non-linear feedback is one of the differences between linear and non-linear equations, that is, non-linear equations have terms which are repeatedly multiplied by themselves. Aslin (1993: 392) uses the following illustrations to explain the relationship between multiplication and non-linearity (see Figure 8).

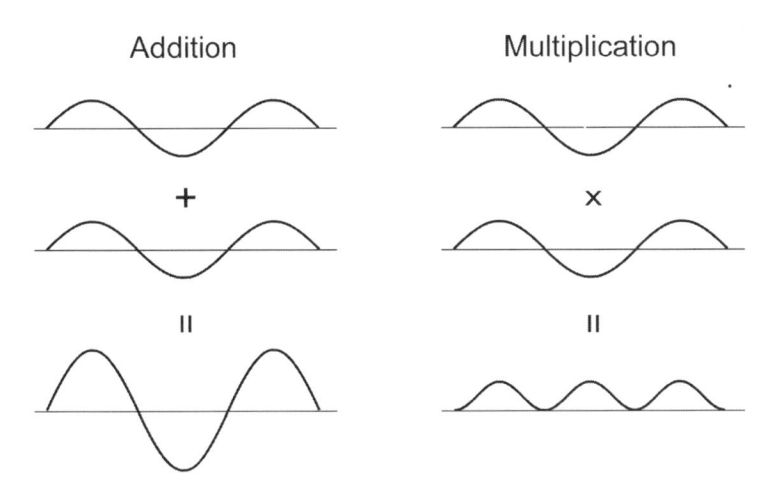

Figure 8 (a) Addition versus (b) multiplication (based on Aslin, 1993: 392)

[T]he manipulation of a single parameter can have a very large (and nonlinear) effect on the state of the system. One way in which this single control parameter can have a nonlinear effect upon the interactions among a set of independent variables is by multiplication (see Figure 8b). As shown in Figure 8a, a point-by-point addition of the two waveforms yields a waveform with larger amplitude but the same general shape. However, as shown in 8b, a point-by-point multiplication (convolution) of the two waveforms yields a qualitatively different waveform. Thus, if we have N components that interact to yield a global behavior, but one or more of these components interact with the others in a multiplicative rather than in an additive manner, then this critical component can render the entire system nonlinear. (Aslin, 1993: 392)

Non-linear feedback can be called 'negative' and 'positive' and the two adjectives simply indicate that one type of feedback regulates and the other amplifies. These two basic kinds of feedback can be found at all levels of living systems, such as in the evolution of ecology and in the psychology of social interaction. Like non-linearity feedback embodies an essential tension between order and chaos since it 'can turn the simplest activity into the complex efflorescence of a fireworks display.' (Briggs & Peat, 1989: 24–25)

Qualitative change in organisation is produced by feedback and characterises non-linear and complex systems, with good examples being agricultural and ecological systems, politico-ecomomic systems, industrial plants, etc. The following definition of qualitative change is suggested by Krohn and Küppers (1992: 392):

> Begriff, der aus der Physik kommt und dort ursprünglich für die Übergänge zwischen fester und flüssiger sowie zwischen flüssiger und gasförmiger Phase verwendet wurde, inzwischen auch auf andere Übergänge [...] übertragen wurde. Von besonderem Interesse waren im letzten Jahrzehnt Phasenübergänge im thermodynamischen Nichtgleichgewicht, zum Beispiel beim Laser und der Bénard-Zelle, aber auch beim Auftreten chemischer Oszillationen sowie in der Biologie, Ökologie, Ökonomie etc. In der mathematischen Beschreibung handelt es sich um einen Wechsel der Stabilität. (The concept originally derived from physics and was used to describe the transitions from liquid to gaseous state, and has in the meantime been transferred to other transitions. Particular interest was paid to the transition phases in the thermodynamic non-equilibrium, e.g. the laser in the Bénard Cell or the occurrence of chemical oscillations as well as in biology, ecology, economy etc. In mathematical terms it can be described as a change in stability. Translation by the authors)

A Dynamic Model of Multilingualism Developed

To describe the complex relationships in dynamic systems Smith and Thelen also draw on the field of thermodynamics, the study of heat transfer and the exchanges of energy and work:

> The cause of all the structures in the following figure and the cause of activity of any one molecule of gas is a single complex system of relationships. Three aspects of this system are noteworthy for developmental theorists. First the system of relationships is complex and nonlinear. Increases in heat do not cause just increases in activity; they cause qualitative shifts in pattern. Equal increments in heat do not cause comparable changes in organisation; rather, at some ranges of heat, large changes do not alter the pattern, whereas at other points on the heat continuum, small changes cause qualitative shifts in the pattern. Second, all the possible structures result from a single system. A single set of equations can be written to describe this system. Third, there are no icons. The system behaves in stable and organised ways, yet nothing in the containers, gas, or heat looks like the emerging pattern. The structures are products of local processes, not the causes of them.

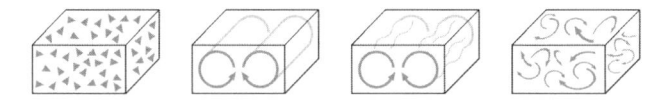

Figure 9 Changes in the movement of gas particles in a chamber as a function of heat (based on Smith & Thelen, 1993: 161–162)

The concept of entropy is also strongly linked to the field of thermodynamics, where it serves as an adjunct of the Second Law, the tendency of any system in the universe to slide toward the state of increasing disorder or heat death. In other words, global systems tend to organise themselves as if they were minimising the energy required to perform a particular class of behaviours: mixing, disorder, randomness. Or in other words: entropy causes open systems to give up their energy to the surrounding environment (see Gleick, 1987: 257; Briggs & Peat, 1989: 27).

Self-organisation and entropy are thus related concepts forming part of the fundamental ideas of chaos theory. Prigogine (e.g. 1980; Nicolis &

Prigogine, 1989) was one of the first scientists to discover that strange things occur in a far-from-equilibrum turbulent chaos where systems do not just break down but new systems emerge. Equilibrum can be defined as the state maximum entropy where molecules are paralysed or move around at random. One of the properties of far-from-equilibrum chaos is that it contains the possibility of self-organisation. Self-organising structures are seen as emerging in biology, in vortices, in the growth of cities and political movements, in the evolution of stars. He calls instances of disequilibrum and self-organisation 'dissipative structures.' This means that in order to evolve and maintain their shape, open systems use up energy and matter. Thus they take in energy from the outside and produce entropy (i.e. waste, randomised energy) which they dissipate into the surrounding environment (see Briggs & Peat, 1989: 136–138).

The Belousov-Zhabotinsky reaction is often used as an example to illustrate the radical new properties which enable self-organisation out of a chaotic background (see Figure 10). In this figure X and Y are the starting chemicals, P and Q are end products of the reaction. The middle shows the autocatalytic feedback iteration sustaining the reaction. Whereas in an ordinary reaction, a system ends up with homogeneous, unstructured mixtures of chemicals, in some reactions, one type of molecule cannot be made unless it finds itself in the presence of its own type. Such a chemical becomes its own catalyst or in other words it iterates. (Briggs & Peat, 1989: 140)

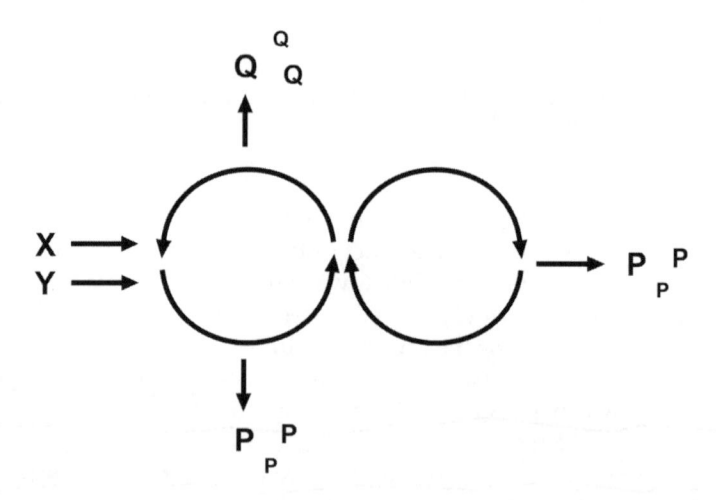

Figure 10 Belousov-Zhabotinsky reaction (based on Briggs & Peat, 1989: 140)

A Dynamic Model of Multilingualism Developed

Chemists call these reactions 'autocatalysis,' because they involve processes in which the products of some steps feed back into their own production or inhibition. Such chemical iterations lead to chemical systems which exhibit everything from equilibrum and limit cycles to period doubling, chaos, intermittency, and self-organisation. The systems structure space by grouping the reacting molecules into orderly patterns of a certain size, and they mark time by evolving and changing constantly. They never maintain the same basic organisation. (Briggs & Peat, 1989: 140)

Living systems show unique characteristics which have been discussed in the concept of autopoiesis or autopoietic structures (e.g. Maturana, 1998). These include constant self-renewal produced by feedback loops in a living organism in contrast to machines. With the help of loops the internal organisation of a living entity is able to adjust to the demands of its environment. Whereas negative feedback loops regulate, positive loops amplify.

Autopoietic systems can be described as 'creatures of paradox', that is, because autopoietic structures are self-renewing, they are highly autonomous, each one having its separate identity, which it continuously maintains. Yet, like other open systems autopoietic structures are also inextricably embedded in and inextricably merged into their environment. Thus each autopoietic structure has a unique history which again is tied to the history of the larger environment and other autopoietic structures leading to an interwovenness of time's arrows. Autopoietic structures have definite boundaries, such as semipermeable membrane; they are open boundaries through which the system is connected with almost unimaginable complexity to its environment. (Briggs & Peat, 1989: 53–54)

In order to study complex systems wholistic concepts of systems in development have become a necessary part of research. This new perspective on reality has been gained with the help of qualitative measurement as used in qualitative mathematics. In contrast to a quantitative mode where it is important to see how one part of the system affects other parts in qualitative measurement the system's movement as a whole is plotted and compared to other whole systems (Briggs & Peat, 1989: 83).

> The scientists of change have learned that the evolution of complex systems can't be followed in causal detail because such systems are holistic: Everything affects everything else. To understand them it's necessary to see into their complexity. Fractal geometry provides that vision: a picture of the qualities of change. Perhaps some of the appeal of the fractal is that in each of its 'parts' it's an image of the whole, an image in the looking glass. (Briggs & Peat, 1989: 110)

At the same time a dynamic systems approach makes clear that in this non-linear world which is holistic everything is interconnected and surface structures can be seen as implicitly correlated to a high degree. This leads to the assumption that there must always be a subtle order present. Briggs & Peat (1989: 127), who base their book on the metaphor of the turbulent mirror, see this order as closely linked to the soliton behaviour (i.e. triggering of the underlying correlations which emerge to shape the system) of the chaotic system:

> On one side of the mirror, the orderly system falls victim to an attracting chaos; on the other, the chaotic system discovers the potentiality in its interactions for an attracting order. On one side, a simple regular system reveals its implicit complexity. On the other, complexity reveals its implicit coherence.

Order, chaos, complexity and wholeness are all tied together and thus nowadays more and more researchers feel the futility of studying parts in isolation from the whole in most of the sciences. Briggs and Peat (1989: 147–148) are very clear on the problems complexity theory has to face.

> Complex systems – both chaotic and orderly ones – are ultimately unanalysable, irreducible into parts, because the parts are constantly being folded into each other by iterations and feedback. Therefore, it is an illusion to speak of isolating a single interaction between two particles and to claim that this interaction can go backward in time. Any interaction takes place in the larger system and the system as a whole is constantly changing, bifurcating, iterating. So the system and all its 'parts' have a direction in time.

Time thus becomes an expression of the system's holistic interaction, and this interaction extends outward. Every complex system is a changing part of a greater whole, a nesting of larger and larger wholes leading eventually to the most complex dynamical system of all, the system that ultimately encompasses whatever we mean by order and chaos – the universe itself.

We have obviously introduced only a few but nevertheless crucial ideas related to dynamic systems approaches. We would like to finish our short introduction with the following illustration based on Briggs and Peat (1989: 176) of the process of making a non-linear feedback model which can be described itself as a non-linear feedback process (Figure 11).

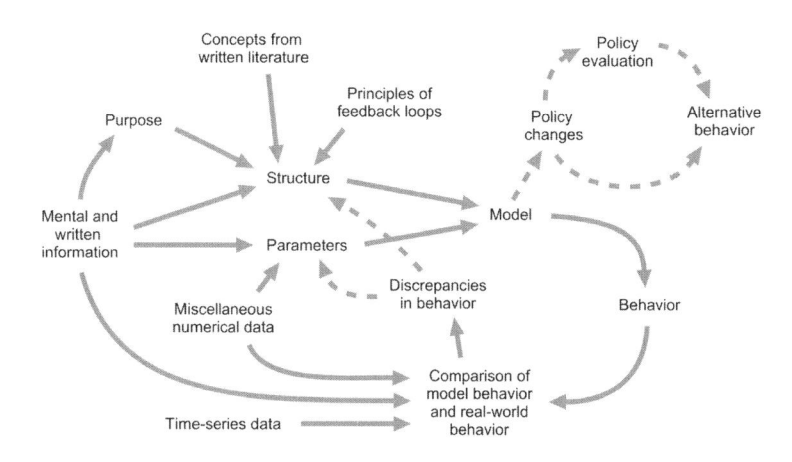

Figure 11 Non-linear model of making a non-linear feedback model

Building DMM

DMM and dynamic systems approaches

Our dynamic model here draws on related fields of knowledge, the most important being systems research, general biology, and cognitive psychology. It is quite obvious that the questions raised by DMM cover a very wide field of neurological and psycholinguistic issues, whose implications cannot be discussed in detail here.

This does not mean that neurological aspects are considered secondary. In fact, recent research on neurological issues does provide an implicit foundation of DMM, though this is not explicated here. Apart from the ACT model developed by Anderson and the declarative/procedural distinction popularised by Ryle (1979) [1948] – but in fact going back at least to St. Augustine – there are many other neurolinguistic insights drawn upon, as to be found in Gardner (1983), Anderson (1983, 1995), Ornstein and Thompson (1984), Vincent (1986), Gigerenzer and Murray (1989), Lurija (1991), Robert (1994), Roth (1994) and many others.

The third strand of research is that of cognitive psychology and cognitive science, where our interest lay not in the empirical findings concerning brain functions, but rather the general arguments concerning the way mental functions can be interpreted.

In contrast to nativist and reductionist approaches where (Bates &

Elman, 1992: 75): '[c]hange – in so far as we see change at all – is attributed to the maturation of predetermined mental content, to the release of preformed material by an environmental 'trigger', and/or to the gradual removal of banal sensory and motor limitations that hid all this complex innate knowledge from view', dynamic systems are to be seen as evolutionary systems that continue to develop in a process of interaction between the dispositions of the system and input from the environment. Dynamic systems are therefore marked by a teleological principle of dynamic homeostasis, a process of constant adjustment to the changing environment and internal conditions aiming at the maintenance of a state of (dynamic) balance. This process of evolution can also be termed 'learning'. Language acquisition theory is seen as a necessary part of the theory of dynamic systems as applied to multilingualism.

Goals of DMM

A dynamic model of multilingualism as presented in this book is meant to achieve the following goals:

- to serve as a bridge between SLA research and MR.

DMM provides a tool with which we can view learner systems and stable multilingual systems as variants of multilingual systems fundamentally obeying the same principles. This makes the traditional distinction between MR and SLA superfluous as both are assumed to obey the same principles.

- to indicate that future language acquisition studies should go beyond studies of the contact between two languages, turning their attention towards trilingualism and other forms of multilingualism.

As pointed out in the previous chapters, traditional contrastive approaches have tended to draw attention away from the aspect of the ongoing development of the multilingual system.

- to overcome the implicit or explicit monolingual bias of MR through the development of an autonomous model of multilingualism.

As long as MR research does not avail itself of an autonomous theoretical basis, not merely relying on the findings of FLA and SLA research, both the results and predictions of research will always be distorted by the assumptions of individual language acquisition research, be it FLA or SLA, as both forms focus on either L1 or L2 but not on the development of LS_1 and LS_2 as interdependent language systems forming part of an overall multicomponential psycholinguistic system.

- to provide a scientific means of predicting multilingual development on the basis of factors assumed to be involved.

So far MR was hampered by the fact that there was no underlying concept of multilingual acquisition as developmental aspects of multilingualism were not a prime object of research. Being able to model the development of multilingualism in time is, however, a necessary prerequisite of a methodologically adequate approach to the phenomenon and research into the individual factors involved.

- to provide a theory of multilingualism with greater explanatory power.

Compared to other models of multilingualism DMM sports the following advantages.

Firstly, DMM provides an explicit model of multilingualism, which means that predictions can be made on the basis of DMM and the model modified to correspond to the empirical data obtained. DMM provides multilingualism with a useful theoretical framework in which the findings of FLA and SLA research can be integrated and which can be extended to cover TLA. DMM provides an explanation of interaction phenomena and both positive and negative language growth, of which the latter has generally been ignored by language acquisition research for a long time.

Secondly, DMM provides a dynamic model of multilingualism. It is not satisfied with the description of the relation between individual factors involved in multilingualism, but attempts to model the development of the multilingual system in time. The authors believe that it is only in the understanding of the dynamics of the multilingual system that multilingual phenomena can really be understood. And we hope that with the help of DMM new research areas will be developed and plausible explanations will be found for hitherto puzzling phenomena such as the last language principle suggested by Shannon (1991), which cannot be explained with the aid of traditional concepts.

Variables and constants

Within a psycholinguistic model we can assume necessary parameters in a systems-theoretic approach, in which we have to distinguish between variables and constants (see, e.g. van Geert, 1994: 325) which form part of phase space.

> Phase space is composed of as many dimensions (or variables) as the scientist needs to describe a system's movement. With a mechanical

> system, scientists usually map the system's phase space in terms of position and velocity. In an ecological system, the phase space might be mapped as the population size of different species. Diagramming the movement of a system's variables in phase space reveals the curious byways of an otherwise hidden reality. ... [N]ature's systems will often undergo rigid, repetitive movements and then, at some critical point, evolve a radical new behavior. It is these changes of behavior that the phase space maps help to clarify. (Briggs & Peat, 1989: 33)

In DMM we assume that the multilingual psycholinguistic system contains factors which can be interpreted as constants. This cannot be taken to imply that these factors are necessarily invariable, but does mean that they function as 'dispositions' and are in effect invariable within the given space of time. These factors may include cognitive capacity, language aptitude and other parameters specified below. These determine the basic form of multilingualism; the constants are at least partially responsible for the development of transitional bilingualism, asymmetrical or unbalanced bilingualism or balanced bilingualism under the same sociolinguistic conditions. To these constants are to be added variables such as perceived language competence, self-esteem, language anxiety and motivation determining language development and language loss.

> The promise of dynamic systems as a theory of development also lies in its connection of real-time processes to change over developmental time. Structural theory views variability in real time as noise and leaves variability in developmental time unexplained. Dynamic systems theory offers a way of thinking about how one might emerge from the other – how the phase space might change as a function of the regions of the phase spaces that have been visited in the past. This idea of developmentally dynamic phase states that are dependent on the history of the system was already proposed by Waddington (1977). (Smith & Thelen, 1993: 763)

The changes in the dynamic systems are not only to be explored in detail but to be illustrated by graphs. Note that all the curves used neither start from complete zero, as a certain, albeit negligible knowledge of the language can be presupposed, nor do the curves reach an absolute level, as the learning curve levels out before complete command of the target language is achieved. We must also note that the graphs simply relate language learning to time needed and predict the modifications in expected language growth due to the effect of certain factors assumed to affect multilinguals and ignore the fact that the level of achievement indicated is heterogeneous even in monolinguals let alone multilinguals. Please also note that in a systems-theoretic approach we base our discussion not on languages (L1, L2, L3, etc.) but on the development of individual language systems (LS; LS_1, LS_2, LS_3, etc.) forming part of the psycholinguistic system of the multilingual speaker.

Characteristics of a dynamic multilingual system

In accordance with dynamic systems theory we wish to identify the following characteristics of language development in multilingual systems:

- non-linearity
- reversibility
- stability
- interdependence
- complexity
- change of quality

In DMM it is assumed that the development of the variables obeys the principles observed in the development of living organisms. As said before, the development of biological systems manifests certain universal principles that do not apply to non-living systems. Biological growth processes are modelled in terms of feedback-loops, sine curves and changes of rates of growth and not in terms of switches, levers and linear relations as presented in reductionist theory (see Chapter 4).

In other words, the theory of language learning adapted here explains the progress made in learning a language, be it first, second or third (etc.) as non-linear in contrast to other approaches where this progress is interpreted as an ordered sequence of individual steps suggesting a steady upward motion where one step follows on the other. Thus traditionally language acquisition has been linked with the idea of a linear process leading to ideal native-speaker competence (Figure 12). Most current theories of language learning presuppose an implicit linear model of development. This graph also expresses the assumption of monotonous or homogeneous growth.

In DMM, however, we assume that according to growth patterns in real life language acquisition is interpreted as non-linear, which means that under ideal circumstances we obtain a growth curve showing some similarity to a sine curve where we can observe slow growth which then increases its rate of acceleration before finally slowing down to achieve a state of equilibrum:

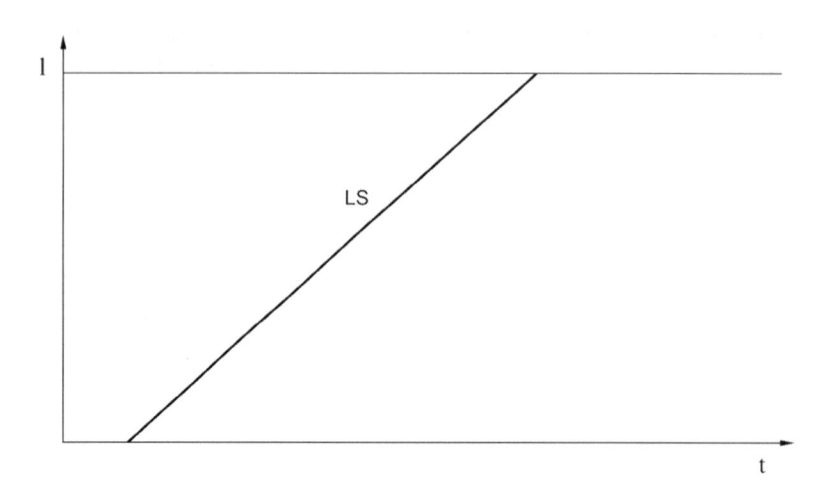

Figure 12 Linear process
LS = language system; t = time; l = language level

> If we look at the actual growth that one is likely to find in a real biological system [...] there is often a short period at the beginning, known as lag phase, in which the system is adapting itself to its surroundings. Then it may grow for a long time exponentially [...] But eventually the rate of growth begins to slacken, the curve of size begins to rise less steeply, and then turns over and becomes flat. (Waddington, 1977: 73)

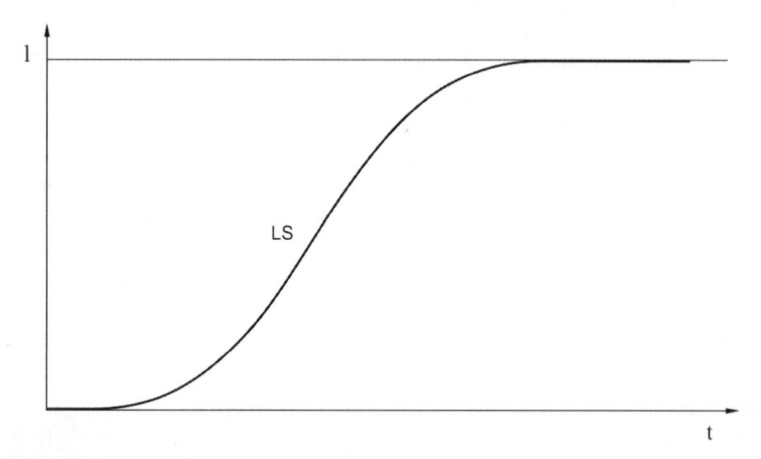

Figure 13 Biological growth
LS = language system; t = time; l = language level

According to biological principles language development is seen as a dynamic process with phases of accelerated growth and retardation. The development is dependent on environmental factors and is indeterminate (Figure 13).

The sine curve has to be interpreted as a rough idealisation of the actual development to be observed in multilingual systems since it can only describe an approximation of the effective development of the system as the acquisition process is characterised by individual stages of improvement and restructuring (see, e.g. McLaughlin & Heredia, 1996). The reasons for the flattening of the curve lies in the limitedness of the learner's resources, that is, the learner has a certain amount of time and energy available to spend on learning and maintaining a language.

We thus argue that in a psycholinguistic context a learner will gradually lose knowledge of a language if not enough time and energy is spent on refreshing the knowledge of an L2 or L3. Thus the theoretical progress in language learning or positive growth to be expected can be counteracted by negative growth which will eventually result in language attrition or gradual language loss, a mirrored process of language acquisition represented by an inverted sine curve (Figure 14).

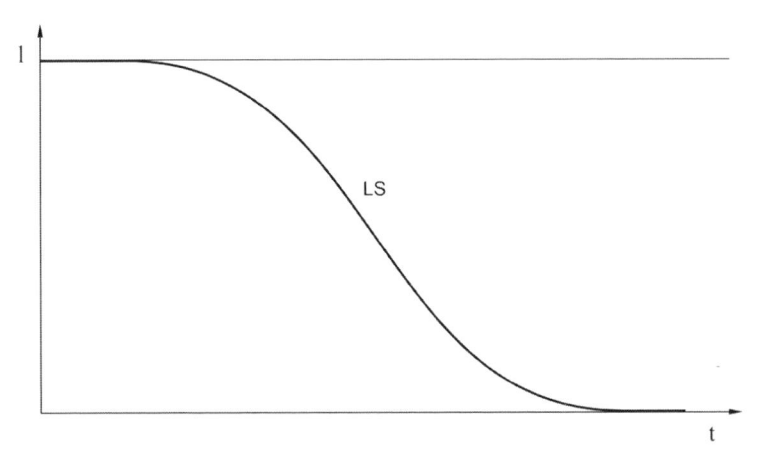

Figure 14 Gradual language loss
LS = language system; t = time; l = language level

Gradual language loss represents an inversion of language growth. Lack of maintenance of a language system results in an adaptive process in which language competence is adjusted to meet the perceived communicative needs of the individual speaker.

The stability of a psycholinguistic system is therefore dependent on the requirements of language maintenance, that is, if the learner does not invest enough energy and time into maintaining a system's stability as a desired effect the system will erode. Obviously systems stability is also dependent on other factors such as the number of languages involved, maturational age at which a certain language is learnt and relative stability established, the level of proficiency at which this takes place and the time span over which the language system is maintained.

A significant element in DMM is the fact that within the psycholinguistic model language systems are seen as interdependent and not as autonomous systems as they are perceived in transfer and CS research. This means that the behaviour of each individual language system in a multilingual system largely depends on the behaviour of previous and subsequent systems and it would therefore not make sense to look at the systems in terms of isolated development. If we assume the validity of the underlying systems-specific parameters in the multilingual speaker, then the subject-specific factors determine both the complexity and variability of the system and on the other hand the given systems are influenced both in their development and structure by crosslinguistic effects.

According to DMM seemingly identical phenomena of transfer and/or interference can lead to divergent results in different systems, even if they are subsequently or transitionally commanded by the same speaker as is the case in transitional bilingualism where one language system is gradually displaced by another one. This paradox of transfer is even less of a paradox, when we take into account the fact that the interfering linguistic subsystems (in a bi-/multilingual) themselves do not represent constants, but are developing (or potentially decaying) subsystems.

Thus we hope to be able to show that DMM will provide the necessary conceptual framework to explain what we refer to as 'the paradox of transfer' as outlined in Chapter 3. For in dynamic systems approaches we are used to encountering systems which in different (aggregate) states will react differently to identical input. In DMM we can distinguish between different language systems commanded by the native speaker (i.e. monolingualism, balanced bilingualism, dominant bilingualism, ambilingualism, etc.) which viewed as (separate) systems exhibit different properties.

The acquisition of several language systems results in a qualitative change in the speaker's psycholinguistic system, that is, as the whole psycholinguistic system adapts to meet new psychological and social requirements, it also changes its nature. Thus the acquisition of a further language leads to the development of new skills forming part of the multilingual repertoire.

Both the overall linguistic systems and the constitutive subsystems can be interpreted as steady states (or point attractors) in systems-theoretic terms and the speaker can be interpreted as shifting from one (relatively) steady linguistic state to the other in the process of language acquisition or language loss.

> 'State' does not mean the same thing as 'stage' or 'phase' (see van Geert 1986). It is a very general term used to denote whatever developmental level concept you wish to employ. For instance, for a continuous developmental function such as a continuous increase in the mastery of a skill, the developmental state could be any point on the continuous curve that models the increase. (van Geert, 1994: 20)

Steady states means that the development stays at the same level, that is its inputs are identical to its outputs. So whereas growth produces change, the steady state merely maintains the system (see van Geert, 1994: 32–33). Thus while one of the languages available within the multilingual system will be maintained or stabilised another might pay the price of insufficient speaker resources and undergo a gradual process of erosion.

Here bilingual studies have highlighted one of the major shortcomings of traditional language acquisition theory, in so far as this focused exclusively on language acquisition, whilst ignoring the process of gradual language loss, which forms the basis of one form of multilingualism known as transitional bilingualism as described above. Indeed, the UG interpretation of language acquisition does not admit of the possibility of language unlearning, but sees the language learning process as unidirectional, an assumption all too easily disproved by empirical evidence (see Hoffmann, 1991: 173). In the following discussion of gradual language loss and language maintenance their interwovenness and importance in multilingual systems as described in DMM will become even more apparent.

Key Factors of DMM: Gradual Language Loss and Language Maintenance

Gradual language loss

In the case of multilinguals we are frequently confronted with the phenomenon of language loss, language deterioration and/or attrition, a phenomenon frequently observed by sociolinguists yet rarely interpreted in a psycholinguistic context. Language loss, which can affect both L1 in an L2 environment and L2 in an L1 environment – leaving aside multilingual situations – has generally been discussed in terms of generation shifts (i.e. the transition within a multilingual community, for example, from active bilingualism to passive bilingualism or reversing language shift) rather than a linguistic experience affecting members of a (multilingual) speech community (see Downes, 1984: 199; de Bot, 1996: 579–585).

> From a psycholinguistic point of view, the pattern language loss takes may offer insights into the structure of the linguistic system. In much the same way as language acquisition is believed to be governed by general principles of language and language ability, patterns of language loss are believed to offer a similar view of language, be it from the other end. [...] [T]he explanation for systematicity in both fields raises similar questions: the universality of the process, the role of interlinguistic versus intralinguistic factors in the explanation of the process, the degree to which competence and/or performance is involved. [...] In another sense, language loss relates closely to research in language variation and language change. (Fase & Jaspaert & Kroon, 1992: 9)

The simplest approach towards the idea of language attrition is the assumption that the forgetting of language underlies the same principles as those affecting general cognitive contents. But one of the problems of research on forgetting is based on the tendency in experimental research to focus on items rather than on systems and we can therefore not be sure whether results obtained from research on forgetting can be transferred to language-related research (see Weltens & Grendel, 1993: 137).

There are doubts expressed in psycholinguistic research whether the forgetting of a language or part of a language is possible at all. According to the common literature on forgetting there is no loss of information from memory, that is, due to numerous factors information can become inaccessible but is retrievable if the right cues are used (de Bot, 1996: 583). We would like to draw the reader's attention to two general theories of forgetting which could be applied to language attrition research. The first is the theory of forgetting as a gradual process of information decay where forgetting is seen as a function in time, that is the longer the phase between learning and forgetting, the more difficult or less likely the particular recall of an item of information will be. The second is a cognitive interference theory where degree of access to information is reduced because old information is covered up by new. Thus the linguistic knowledge relevant to two (or more) language systems is interpreted as competing for memory space and recall as known from cognitive linguistics research by MacWhinney and Bates (1989).

Indeed, subjective or individual language loss is a well-known phenomenon and not merely to neurolinguists interested in various types of aphasia. The fact that language loss can affect healthy and proficient adult speakers and take effect in a comparatively short period of time (without changes in other cognitive abilities) is illustrated by the following example quoted by Aitchison (1981: 179):

> [I]t is clear that people can and do alter their speech quite considerably in their adult life, as is seen from the case of people who emigrate, or who move to a socially prestigious area, and adjust both their accent and sentence structure to those of their neighbours. Sometimes the adjustment to those around is quite rapid, as shown by an English speaker, recently imprisoned in Delhi: 'Mary Ellen Eather, the Australian, was led through her statement by the prosecution lawyer. After a year in jail she now spoke with an Indian accent [...]. (Neville & Clarke 1979)

It may therefore be suggested that the problem of language unlearning observed in child language acquisition (see Pinker, 1989: 290) has its equivalent in language loss observed in adult speakers. Generally, however, language loss has been investigated in terms of language death under pressure of a competing or dominant language, that is, in a fundamentally bilingual situation. This might create the inaccurate impression that language loss or unlearning only takes place in linguistic situations where languages are in sociolinguistic and therefore frequently also psycholinguistic competition. We may rather assume that language loss can take place in normal and healthy monolingual speakers as well, that is in the form of intrapersonal (systematic) variation.

One reason why language loss still tends to be ignored is because the phenomenon has long been misunderstood. Perhaps influenced by early work on aphasia and traumatically induced, that is pathological (complete) language loss, it tended to be viewed in terms of a sudden neurophysiologically determined loss of performance or even competence. As Seliger (1996: 616) states: 'Interestingly, explaining the development of primary language attrition is even more problematical. In cases other than language pathology, we do not expect an established L1 to deteriorate or diverge from the grammar that has been fully acquired.'

It has therefore become customary to distinguish between language loss and attrition, with a tendency to focus on loss whilst largely neglecting attrition. From a systems-theoretic point of view so-called 'complete' language loss due to external influence on the system is of little interest as it does not elucidate systems behaviour. So for our purposes here the standard form of loss must be 'language attrition,' which in a systems model is consistently called gradual loss.

Why has language attrition generally been ignored by research?

- Language attrition is a gradual and much less spectacular process than abrupt complete language loss.
- If noticeable, the speaker will generally try to counteract the effect of language attrition by increased compensatory strategies (see Turian & Altenberg, 1991). But there are cases where L1 is not considered prestigious enough to be maintained as found, for example, in German speakers who immigrated to the US for political reasons.
- Language attrition is not observable because (according to DMM), at least at an early stage, it expresses itself in the form of an increased scatter of performance. As long as there is no explicit performance measure, this increased scatter will go unnoticed. But we do find comments on variability observed in language attrition in Altenberg (1991).

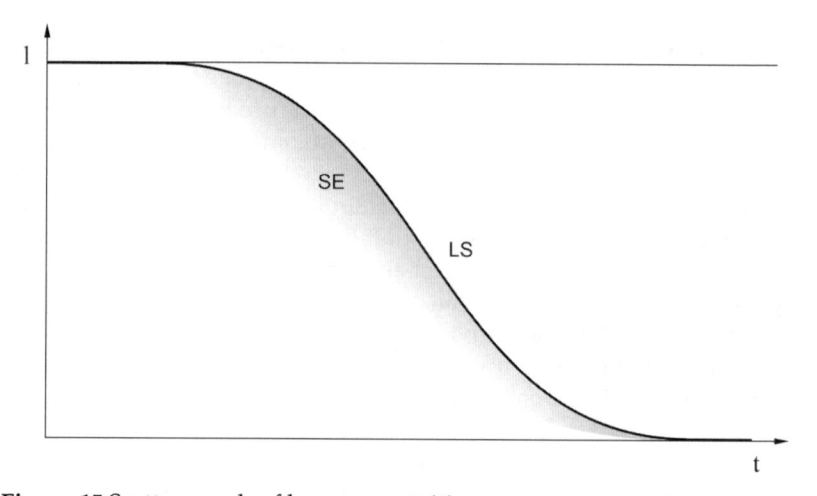

Figure 15 Scatter graph of language attrition

SE = scatter effect; LS = language system; t = time; l = language level

Figure 15 shows performance scatter in the course of gradual language loss. Increased scatter indicates the erosion of proficiency prior to negative growth of the language system. Scatter is reduced as soon as the language system attains a new level of stability and it is specified as negative as it is always only observed as deviation from language norms. There is, therefore, by definition no positive scatter.

Within a mathematical model we might see those points at which the systems retains a new level of stability as basin of attraction. In a comparison of a dynamic system to a topological landscape a basin of attraction is represented by a valley where a system comes to rest (see van Geert, 1994: 58). The guiding principle must, however, be the idea of dynamic homeostasis. It would be easy to misunderstand this state of equilibrum as an invariable steady state, but within a living system we must assume that homeostasis is essentially dynamic as maturational changes will counteract any tendencies towards final stabilisation and therefore rather represents a ball inbetween constantly moving goalposts.

According to DMM easily perceived forms of language loss will follow a phase of declining language use which appears to have no effect on a language system (see Weltens, 1988: 711). For instance in the case of ethnolinguistic minority children it is often difficult to know whether children's limited competence in L1 during later childhood and adulthood is better characterised as language attrition or as a phenomenon called language arrest by Pan and Berko Gleason (1986: 196). As argued below, it is probably more realistic to adopt an ecological view of the declining and/or threatened language system. We can therefore expect the system to attempt to absorb the effect of decreasing language use by internal adjustments which are not perceived by the outward observer. As argued by Harley (1994: 689), this kind of language loss is likely to be delayed, as on the face of it, lack of use does not have any effect in well-developed language systems:

> The level of proficiency that learners reach has emerged as one key indicator of the future level of retained proficiency (Bahrick, 1984; Weltens, van Els & Schils, 1989). Indeed a group of researchers in the Netherlands found only a small amount of attrition among high proficiency learners of French, even after several years away from their high school classes and with little or no subsequent exposure to the second language. (Weltens *et al.*, 1989).

A critical stage model is probably an adequate representation of what happens in the threatened language system. We know that language systems are necessarily redundant and, for example, a loss of synonyms or quasi-synonyms in the lexical system will represent a systems adjustment that will generally go unnoticed in everyday language use as everyday communicative situations do not require the use of a large number of synonymous expressions (see Aitchison, 1981: 186). Note also that language loss will affect different linguistic subsystems to an unequal degree. DMM also assumes that the speaker can compensate for a shrinking language system by increased monitoring effort in the course of speech production and interpretation (see also Chapter 7). It has been noted that weak or underdeveloped systems have less scope for absorption of the effects of lack of use, which means that these effects are more clearly visible in what DMM terms partial systems.

It is also important to note that measures of language loss are necessarily unreliable as it is extremely difficult to measure language use – that is, exposure is as difficult to trace as to avoid. If we assume the existence of CLIN, lack of exposure to L1 does not necessarily mean that components of this language system are not activated (see Chapter 5). Thus we may also assume that one reason why absolute language loss is generally not observed in native speakers is because the tendency towards a state of linguistic entropy is counteracted by the – frequently subconscious – process of language maintenance.

Language maintenance

> This association between maintenance and current use suggests that finding opportunities for communicative practice is an important part of keeping up these skills, or of remedying them when they appear to have receded. New exposure to the language has in fact been observed to lead to a rapid recovery of skills (Clark & Jorden, 1984) – a clear indication of attrition rather than loss. (Harley, 1994: 691)

Being a side product of day-to-day communication, language maintenance is, however, generally ignored by psycholinguists, whilst all speakers are likely to be able to recall some of the language maintenance work conducted and perceived as natural such as looking up the spelling of a word or reflecting on the systematicity of certain grammatical aspects of one's L1, or asking a fellow native speaker about the appropriacy of punctuation and the simple use of one's verbal and lexical repertoire. According to Harley (1994: 708) these mainly metacognitive strategies to retain a language can be broadly categorised in the same way as language learning strategies. But most important is the simple fact that use of language counteracts language loss or decay. This was also stated by Hyltenstam and Stroud (1996: 568): 'The notion of language maintenance is meaningful only in relation to its sister concept of language loss.' (see also Pan & Berko Gleason, 1986: 197). They reserve the term 'language maintenance' for social aspects, that is for situations where a speech community tries to continue to use its traditional language although threatened by language shift to the dominant language of the community and suggest that the term 'language retention' is used to refer to the individual ability to keep up a language system (1996: 567), a suggestion we do not follow here.

Language maintenance effort (henceforth LME) can be seen as composed of:

(1) language use factor, that is, activation of parts of the linguistic system for communicative purposes resulting in a renewal of parts of the subsystem, and of
(2) linguistic hypotheses verification or corroboration factor, that is, the renewal of parts of the speaker's (explicit knowledge of a) linguistic subsystem by means of a verification of hypotheses concerning the language system.

The absence of this kind of language maintenance resulting from lack of use, that is communication in one specific language, will lead to the deterioration of linguistic competence in the respective language. This tendency is only increased by the presence of a second language competing for a position in the speaker's psycho-communicative system (see Aitchison, 1981: 186).

According to this model of LME, which can be seen as an essential part of the systems model, language learning thus requires an increase in language work. This is here not termed maintenance as it includes the practice of new structures and the extension of basic vocabulary to exceed the necessary maintenance effort required to counteract language deterioration determined by the principle of linguistic entropy. According to the systems model it is also self-evident that the amount of language maintenance required to guarantee homeostasis within a linguistic system (i.e. a dynamic steady state), will increase with the accumulation of linguistic knowledge (see Fase & Jaspaert & Kroon, 1992: 74).

As can be seen from the graphs below, the increase in required language maintenance will always exceed the increase of language knowledge to be observed in the learner system. It is therefore obvious that the individual's language maintenance factor provides a natural limit to the language learner's acquisition efforts. This fact not only provides an alternative, less biologistic explanation of the critical period hypothesis, as learning effort is increasingly hampered by language maintenance work, but also provides an explanation as to why the linear conception of language learning is fundamentally flawed. Chomsky's the paucity-of-the-stimulus argument is largely based on an inductive conception of language learning suggesting that it takes the form of linear progress, a plausible assumption in automata, but highly unlikely in biological systems (see Herdina, 1988).

If, however, we assume, as Maturana and Varela (1987) do, that learning is an autocatalytic or autopoietic process then obviously a minimum amount of input can lead to a complexity of consequences that appear to be entirely unrelated to the input itself. The consequences must be seen as a result of interaction between the individual system and the environment. The seeming lack of correlation between language input and language development can therefore not be accepted as conclusive proof that language competence is innate.

Let us now look at the growth types we may expect in the language learning processes involved in more detail. The traditional view of language growth as a linear process has already been described in Figure 12. According to our conception the abstract learning curve underlying switch theory or the instantaneous language acquisition hypothesis originally suggested by Chomsky is to be replaced by an idealised abstract language learning process taking the form of exponential growth. The gradient of the curve must be seen as an expression of the rate of acceleration or rate of deceleration of language system development. We can interpret these rates as expressing a momentum of change, or an inverse expression of stability, that is 'downward' rates of change are inherently self-augmenting and can therefore lead to the complete loss of command of a language. Upward rates of change are also self-augmenting but will rapidly be confronted with the limits to cognitive and time- and effort-related reorganisation as the balance changes in favour of maintenance rather than acquisition.

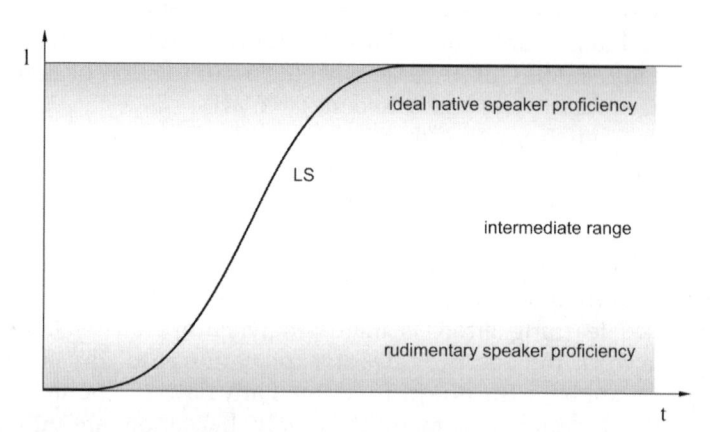

Figure 16 Abstract learning curve without any limiting factors

LS = language system; ISP = ideal native speaker proficiency; RSP = rudimentary speaker proficiency; t = time; l = language level

The graph (Figure 16) provides an unrealistic growth curve outlining language acquisition progress mapped onto proficiency levels and intermediate range.

This type of language growth, which is also used to illustrate the non-linear conception of biological growth (Figure 13), is abstract as it does not take into account those factors slowing down language growth such as interference and language maintenance.

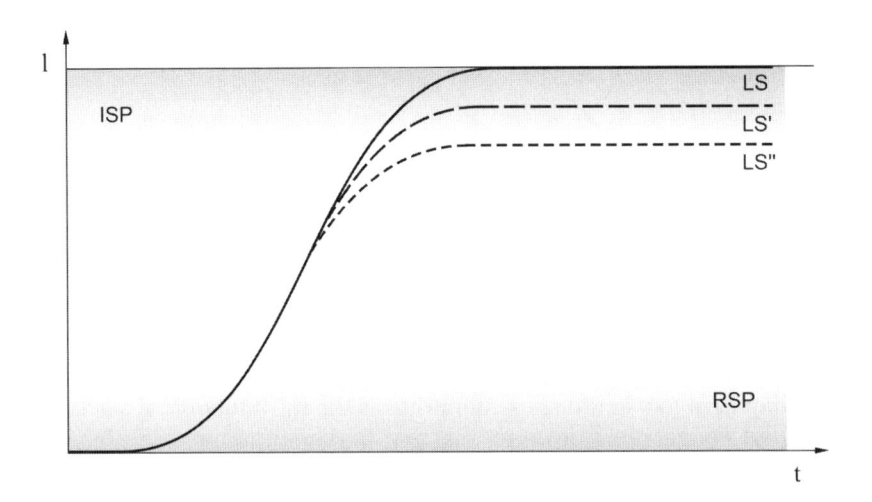

Figure 17 Actual learning curve influenced by limiting factors

LS = language system development without limiting factors; LS' = language system development retarded by insufficient LME; LS" = language system development retarded by insufficient LME and effects; ISP = ideal native speaker proficiency; RSP = rudimentary speaker proficiency; t = time; l = language level

Our model (Figure 17) does not regard the absolute command of a language as a realistic perspective. A certain degree of underachievement is to be expected, influenced by personal limiting factors. This curve is therefore to be attributed to the fact that exponential language growth would require second order growth of language maintenance, something which is soon likely to exhaust the speaker's resources of time, etc. The growth of language systems leads to an accompanying growth in the LME required for the maintenance of the system on the one hand and for language acquisition on the other (see Figure 18).

The term second order growth might at first sight appear confusing, does, however, describe a fairly simple fact. If we assume that language

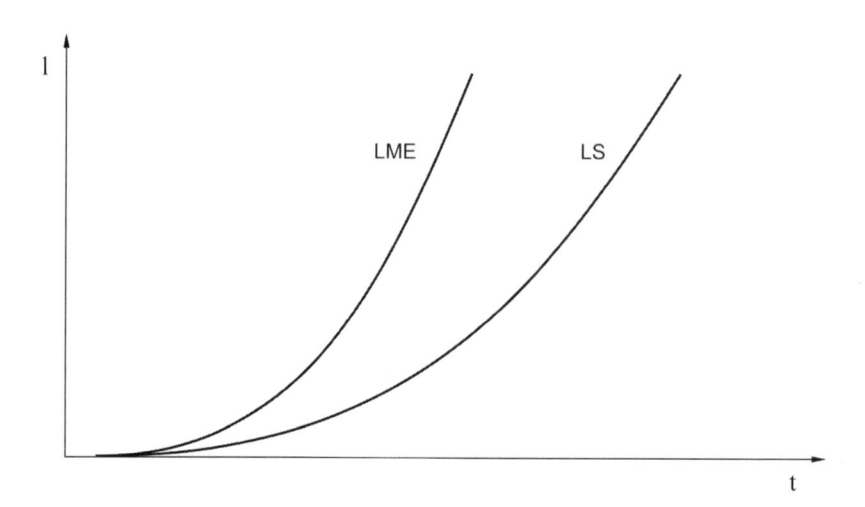

Figure 18 Growth of LME

LS = language system; LME = language maintenance effort; t = time; l = language level

growth at a certain stage is exponential, we will soon realise that the addition of one item to a language system causes the introduction of a greater number of connections between existing and new items. Waddington (1977: 69) explains second order exponential growth as follows: 'One sort of second order growth would be if the fraction of the existing system added on after each interval itself increased, as for instance in a human population in which health conditions were improving so that more of the babies born survived.'

It is, furthermore, important to realise that the processes of language learning and gradual language loss do not take place in an abstract psycholinguistic vacuum but in a concrete sociolinguistic and economic environment, which forces the individual to comply with the principle of economy of effort or least effort. Although similar principles are found in sociology and game-theory the principle of economy of effort can be derived from a generalisation of Zipf's law of least effort (see Zipf, 1968). In view of findings relating to the effort of maintaining a bilingual language system (see Romaine, 1989: 18) we must therefore assume that balanced bilingualism or ambilingualism is most unlikely to remain a steady state in the speaker's system.

Here we can introduce the basic distinction between bilingualism in a monolingual community and multilingualism in a multilingual community. If bilingualism is the result of a transposition from a monolingual society (L1) to a monolingual society (L2), bilingualism is likely to be a transitional state (see Figure 25) not only in sociolinguistic but als in psycholinguistic terms. We can assume that a multilingual speaker (of any category) will always be under pressure to optimise her or his linguistic system in order to achieve greater communicative efficiency. We can therefore establish the principle of theoretical monolingualism (in a monolingual society), implying that despite the frequency of multilingual phenomena, monolingualism is the natural state of a speaker. Just as we can hypothesise that living systems underly the natural principle of entropy, we can assume that the individual speaker in isolation has a natural tendency towards monolingualism, thus minimising LME (see also Seliger, 1996: 617 on the idea of cognitive burden). Therefore bi- or multilingualism must be a result of contrary social or psychological pressures leading to the adoption of a more cumbersome multilingual communicative system, which is, however, constantly under threat from outside (see Downes, 1984).

The frequently observed and much researched phenomenon of switching codes thus contains a new interpretation, as it can be taken as a possible indication of the erosion of both or one of the competing language systems (in case of dominant or balanced bilingualism). That is, CS might be employed because a single system in isolation no longer suffices to cover the communicative needs of the individual and therefore the second(ary) LS has to help out. CS is also indicative of a blurring of the borders between the two LSs as part of LS_2 is, for example, considered functionally equivalent to LS_1. Note that CS is obviously only a useful strategy in a multilingual speaker community, that is the previously mentioned alternative of a bilingual speaker living in a bilingual community, which is likely to lead to a more stable bilingual psycholinguistic system.

This does, however, not exclude the possibility of one of the two languages becoming more dominant than the other, that is the speaker developing a primary (L_p) and secondary language (L_s), which does not necessarily coincide with LS_1 and LS_2. Transitional bilingualism is the simplest form of multilingualism as in an ideal situation LS_1 is simply gradually replaced by LS_2 with all other factors remaining largely constant. Note that such a transition can also occur between LS_2 and LS_3 etc. and vice versa. The complete loss of LS_1 is empirically highly unlikely. In our model such a development can be interpreted as systems displacement resulting from a reorganisation of the storage and accessibility of linguistic information. As suggested above, we assume that the multilingual brain works according to a principle of recall-determined stacking, that is relevant information becomes more accessible depending on the degree of frequency of a recall process (see Anderson, 1995: 198). This hypothesis says little about whether the information relating to L1 is actually forgotten or simply no longer accessible. And the number of studies of re-acquisition or re-emergence of languages in multilinguals (Faingold, 1999) is still too limited to be considered conclusive.

Within the systems model we can indeed predict that the process of language attrition or erosion of the system underlying language competence is more likely to affect less well-educated and/or less communicatively oriented speakers as their LME will be insufficient to sustain a balanced dual language system. Again this factor does not follow simple additive principles as the interaction (i.e. transfer, interference, etc.) between the two systems is going to create further demands on the LME, which should progress geometrically rather than arithmetically. We can postulate critical levels of language maintenance for various types of bilingualism or multilingualism below which attrition of one or other of the language systems is likely to occur, leading to the transition to another linguistic steady state, that is a different form of multilingualism or monolingualism (see Chapter 7).

Thus there are various outcomes of this process, ranging from the reversion to natural monolingualism to the creation of a new language (see Aitchison, 1981: 202–204). The reversion to monolingualism can be observed in Gaelic and Ladin language communities, where we can note individual (and subsequently collective) transitions from ambilingualism to passive bilingualism and finally monolingualism – Irish probably having undergone similar processes – where it is unlikely for an individual to pass through all the stages mentioned as the transition is generally restricted to two or less (see Dorian 1973, 1978; see also Haugen, 1969 on a model of language shift). This limitation is due to the fact that the speaker's sociolinguistic environment has a stabilising effect on the bi- or multilingual system counteracting the principle of subjective linguistic entropy as sociolinguistic change is necessarily more gradual than individual linguistic change (see Chapter 7 on possible determinants of positive and negative growth).

This stabilising influence on the principle of theoretical monolingualism (increasing the LME of the individual) can take various forms, the most unlikely being the existence of a fully multilingual or bilingual, that is ambilingual, community. In this instance both languages would be taken to cover all domains of life and all communicative functions equally well and any member of the speech community could thus be addressed in either language without any resulting differences (see Denison, 1972: 70).

As, however, languages rarely cover all domains of life equally and speech communities are seldom that homogeneous, we are likely to come across a multilingual phenomenon called domain specificity (see Downes, 1984: 49; Fishman, 1972), which will probably give rise to a functional differentiation within the multilingual language system which may be topic-specific, situation-specific or group-specific (see Gumperz, 1972; Romaine, 1989: 151). The introduction of domain specificity will of course contribute to the stabilisation of the multilingual language system as the languages are no longer functionally equivalent and interchangeable but fulfil a complementary role in the daily communicative requirements of the multilingual individual in her/his society (see Hoffmann, 1991: 179).

From a systems-theoretic and cognitive point of view developing domain specific language skills consistitutes a substantial reduction in required LME. The main argument here must be that the reduction is not directly proportional but can be seen as increasing geometrically. The implementation of linguistic division of labour and therefore the creation of a stable subsystem, rather than an unstable fully-fledged all purpose language system, not only has a drastic effect on the amount of maintenance required for a specific language system but also reduces the risk of interference to a minimum.

Similar pressure can be exerted by the existence of a large group of monolinguals with which communication is for some reason or other – economic or ethnic, etc. – a necessity. Whilst the functional differentiation outlined above can obviously stabilise the multilingual system – although it is no longer balanced and by definition will exhibit characteristics of a partial system – a sufficiently influential monolingual group is likely to give rise to transitional multi- or bilingualism leading to the death of the minority language (see Hoffmann, 1991: 41).

According to the principle of minimisation of effort we would therefore expect idealised language attrition to take the form of an inverted sine curve, that is a gradual loss with increasing decline until the language system stabilises at a lower level as already shown in Figure 14. If LME in a specific LS falls below a critical level, the LS begins to erode and to lose an increasingly larger number of elements such as synonymous expressions and complex syntactic structures at an early stage. One of the reasons for the sharp decline after a more gradual initial deterioration can be found in the self-reinforcing processes entailed in language loss: for example, loss of language competence will lead to a reduction in use, as the command of the language is more difficult to maintain and the risk of exposure by goofing and resulting stigmatisation is greater and naturally avoided. The levelling out of the gradual loss curve at a lower level is in part due to the fact that the more rudimentary the remaining LS, the smaller the LME required to achieve a steady state, bearing in mind that language maintenance relates inversely to language loss in second order terms. It is therefore not surprising to find rudimentary linguistic competence being retained over years with minimal LME required.

Summarising we wish to describe LME as dependent on at least two factors:

(1) the language use factor and
(2) the language awareness factor, a factorial specification of what has been discussed as metalinguistic awareness.

Language use is primarily dependent on the communicative environment and the resulting frequency of communicative exchanges in a specific language. Every instance of use of a particular language system constitutes an activation of a particular item or number of items of a language system and thus functions as a memory refresher cycle for a specific LS. Language use therefore has an activating or refresher function contributing to the maintenance of a language system.

As used here, language awareness adds a cognitive component to LME. It is assumed that conscious manipulation of and reflection on (the systematicity of) a language is part of LME. LME can therefore be defined as a function of a language use factor and a language awareness factor. This interpretation of language use is confirmed by neurolinguistic evidence on memory functions (see Wessels, 1982).

In conclusion it has thus to be noted that neither language acquisition nor language attrition can adequately be understood as language learning processes in isolation as is still commonly done in the literature. Also, Sharwood Smith (1989: 188) already pointed out that 'loss should be regarded as an integral part of language acquisition studies as a whole.' As will be analysed in more detail in the next chapter, they have to be seen as an integrated part of an evolving dynamic system where language attrition is defined as a function of language acquisition, with language maintenance providing the necessary link between the two processes and thus making clear that the two processes depend to a much higher degree on each other than has been suggested so far.

Predictions of DMM: disadvantages and advantages of multilinguals

DMM therefore expects multilinguals to be worse speakers of the respective languages than monolingual speakers with the same educational background and the same LME. It is, for example, generally agreed that multi- or bilinguals have a harder time learning two languages and are disadvantaged in purely linguistic terms (see Hansegård, 1975: 129).

> No matter what the conditions of multilingual language learning, it seems to be cognitively more difficult to become bilingual than to become monolingual. Parents often report that their children who are exposed to two languages avoid the problem by refusing to speak one of the languages. It seems that only when the status of the two languages is both high and relatively equal, and when both languages are spoken by individuals important to the child that the child rises to the challenge of becoming bilingual. Children who become bilingual early do so in response to environmental demands, otherwise the language with less status does not develop far. (Ben-Zeev, 1977: 39)

The arguments, however, tend to focus on insufficient language maintenance. Another reason already stated is that in addition to insufficient maintenance we can expect the phenomenon of CLIN between the two (or more) language systems to lead to a retardation in the multilingual acquisition process and a disruption of performance in the application of one or either of the two language systems, which according to the insufficient maintenance hypothesis should be underdeveloped in the first place. Although the effects attributed to CLIN in this context are certainly to be expected, it would be wrong to attribute to CLIN too negative an influence on language acquisition and performance, as within the enhanced monitor conception already mentioned, CLIN must also be seen as the major factor in initiating the autocatalytic developments observed in multilingual speakers as already outlined in Chapter 3.

Another reason why multilingual proficiency can be expected to lie below that of monolingual speakers is found in the fact that communicative efficiency, particularly in a restricted number of domains, is not coextensive with syntactic and semantic well-formedness. This means that under certain conditions a linguistic system will work effectively, although it is only an approximative system of the target language, and the sociopsychological disadvantages of deviant speech are either considered unavoidable or not sufficiently grave to require some remedy (see Richards, 1974). This does not mean that there is no pressure to conform (see Kramarae, 1982), although excessive conformity might be stigmatised.

As was already suggested in Chapter 6, multilingualism is not only hampered by disadvantages – the interrelation between the two or more language systems of a multilingual not being reducible to various types of interference phenomena – as a multilingual speaker is likely to be able to draw on a fund of common linguistic experience on the one hand and on the other hand, s/he develops certain types of linguistic and cognitive skills which are either not present or less developed in monolingual speakers. Before presenting a systematisation of these effects, we can look at the suggestions concerning multilingual advantages made by MR.

The implications of greater language awareness for language competence have still to be clarified and we can therefore presently not make any predictions at this early stage of research. Obviously the multilingual native speaker, confronted with competing linguistic systems early on in her/his life and therefore having a standard of comparison will have a greater awareness of the nature of a language system, be more able to abstract from the particular linguistic forms and may even have a better understanding of the peculiarities of the language s/he is a speaker of (see also Chomsky, 1986: 27 on Orwell's problem).

These functions are conscious functions to be contrasted with the unconscious processes involved in (second) language acquisition. It has been argued that the bilingual speaker requires some kind of monitor governing the selection of appropriate language in a certain sociolinguistic context as discussed under the topic of domain specificity above. The original multilingual switch theory, isolating one language from the other, has been replaced by the conception of a conscious monitoring process determining the selection of the appropriate code or governing CS procedures (see Romaine, 1989: 91). A side effect of the process of language management is the development of greater cognitive flexibility and creativity.

As argued in this book, multilingualism also results in systems-specific properties, which in psycholinguistic terms means that the multilingual speaker will develop skills or abilities not found in monolingual speakers. These emergent properties (for more details see Chapter 8) are at least in part most easily identified with the enhanced metalinguistic and metacognitive abilities leading to more efficient language acquisition.

Both multilingual advantage and multilingual disadvantage are modelled within DMM, specifying as many clearly identifiable relations as possible. Some of the factors introduced are necessarily based on conjecture rather than empirical evidence. This should, however, not impair the validity of the concept and the plausibility of the dynamic view, which will obviously allow for adjustments in the number and significance of factors involved.

Hypothetical Assumptions of DMM

Before we investigate further the relations involved in language acquisition in multilinguals, we would like to suggest the following working hypotheses, or axioms, on which the model is based and which are to some extent necessarily idealisations or ideal typical. The assumptions made concern the topics of multilingual proficiency, language growth and language typology.

Homogeneous multilingual proficiency assumption

The homogeneous multilingual proficiency assumption must be seen as an attempt to deal with a rather complex issue. Although it is custom to speak of a speaker having command of a language L1, the chapter on variability has shown us (1) that this command can vary extensively and (2) that LS_1 has to be taken to consist of a number of subsystems. On the one hand we know that the respective LS_1 is seen as consisting of a syntactic system, a lexical system, a phonetic system, etc., and on the other we know that, for example, in the case of diglossia a speaker of a language will have developed subsystems of the lexical system, syntactic system, phonetic system, etc. A proficiency measure of LS_1 and LS_2 will therefore be a cumulative measure of these various subcomponents. This does not exclude the possibility of, for example, developing a lexical variety measure in LS_1 irrespective of the number of glosses involved. Note that this concept differs substantially from the invariable competence assumption discussed in Chapter 4.

For those who are not adherents of UG, it is a commonplace that in the development of a monolingual system the various skills contributing to language ability do not necessarily contribute to a uniform whole, particularly not to some level of proficiency that can be measured in uniform fashion. It is probably more realistic to assume the existence of a weakly defined bundle of skills that can be measured individually (see Chapter 5). The mean of these individual measures is most likely to correspond to what is shown on the proficiency curves outlined in DMM.

Homogeneous growth assumption

A further idealisation used in DMM is that of homogeneous growth. As argued by van Geert (1994: 82–83), homogeneous language growth curves are necessarily idealisations not corresponding to the empirical data obtained. Homogeneous growth is not to be observed in the data obtained from language growth observation. Language growth rather takes place in stops and starts, sudden spurts of language development and well-documented setbacks (see, e.g. overgeneralisation in child language acquisition). The smooth language growth curve is therefore necessarily an abstraction or idealisation but does in its general trend conform to the empirical data.

Equidistance assumption

One obvious problem discovered by CAH is a necessary result of viewing transfer phenomena as a function of the similarity or difference of the two or more language systems involved. In order to be able to treat all language systems in the same fashion we have to assume that all natural languages are basically identical. In order to be able to make consistent predictions in language acquisition theory we furthermore have to assume that the types of languages involved do not affect the workings of the model, that is, we have to be able to assume that any two or more languages are equally similiar/dissimilar to each other. This is most obviously a hypothesis we can generously describe as a counterfactual 'idealisation.' Yet, it does retain a certain possibility, if we assume that degrees of perceived similarity and/or difference cancel each other out. This must be taken to mean that overtly similar language systems will encourage transfer in the acquisition process but will also suffer from interference due to a high degree of systems similiarity (see Chapter 2). The hypothesis of encouraged transfer is supported by Sharwood Smith (1989: 195) who states that '[c]rosslinguistic support may enhance transferability in that two languages may possess something in common, making it more transferable into a third language.'

Equicomplexity assumption

The second simplifying hypothesis on the topic of language typology to be introduced is the assumption that all languages are considered to be of equal complexity, that is, any natural language is viewed to be equally complex. This again represents a counterfactual idealisation as some languages are definitely easier to learn than others (although such a statement obviously has to be related to the L1 on the basis of which this judgement is made). Perhaps we are stating the obvious, when suggesting that Dutch may be an easy language to learn for a German or English speaker, but is not necessarily easier than German for a native speaker of Japanese. Ignoring the existence of partial systems (see utility systems and pidgins discussed above) DMM assumes that all language systems are of identical degrees of complexity. As yet complexity measures of language systems do not exist and the tradition of proving one's native tongue to be the most suited or best has not provided conclusive results (see Eco, 1994) this appears to be a necessary working hypothesis.

Chapter 7
A Dynamic Model of Multilingualism Analysed

This chapter is intended to describe the complexity of the factors involved in multilingual acquisition (and loss) and their presentation within a dynamic model. After the presentation of our conceptual outline where we will introduce parameters such as LME step by step to finally present the reader with a suggestion of a selection of multilingual types according to a systems-theoretic aproach we will concentrate on the relations between the relevant factors of DMM. We will present our ideas of the components of multilingual proficiency such as the M-factor and put them in relation to what we mean by general language effort. The stability of a multilingual system will then be viewed in terms of language distance and variation in multilingual systems. Arising positive and negative growth patterns are determined by the communicative needs of a speaker which will also be focused on in our discussion. We will end this chapter with a presentation of our ideas of a dynamic concept of multilingual systems by focusing on some of the individual factors in language development. The problem presented by the modelling of autocatalytic changes, which concerns one of the most important suggestions made by DMM, will finally be addressed in this chapter.

Conceptual Outline of DMM

DMM is intended to provide an explanatory framework for threshold phenomena observed in multilingual speakers. We shall not do this by the introduction of a complete systems-theoretic model of multilingual proficiency, but rather the gradual introduction of parameters we can take as influencing multilingual language development. For practical reasons we shall presently not distinguish between the two or more competing linguistic systems (of the multilingual speaker) but treat the multilingual system as homogeneous as discussed before.

As described in Chapters 2 and 3 the threshold hypothesis was originally presented by Cummins with regard to bilingual education. The threshold in his theory refers to a level of competence that a bilingual

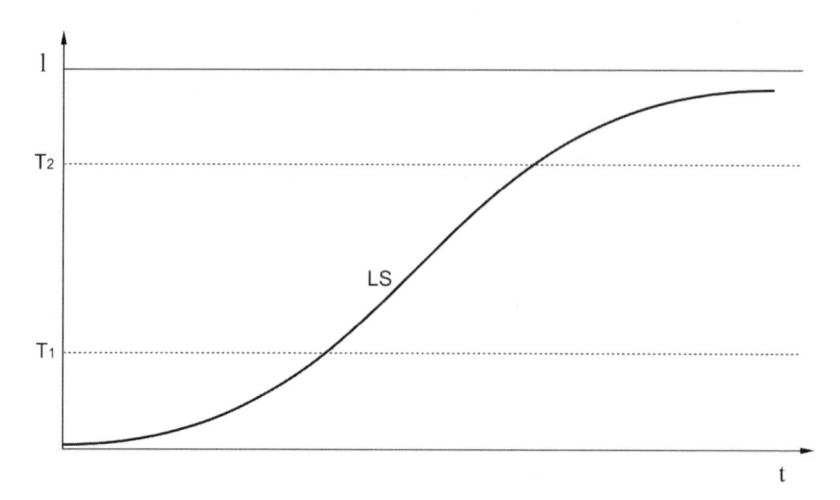

Figure 19 Threshold hypothesis

LS = language system; T_1 = first threshold; T_2 = second threshold; t = time; l = language level

person has to reach in order to gain cognitive benefits from the use of two languages. The various stages of negative, indifferent and positive influence can be interpreted to represent qualitative changes in multilingual development and are therefore of interest to the approach presented here. The graphic representation of the threshold hypothesis in Figure 19 shows the underlying growth curve leading from the phase of negative influence, to indifferent influence and finally positive influence.

In a systems-theoretic model thresholds play an essential role since they indicate a change from one state to another, that is, when a certain threshold value is passed a qualitative change is introduced to a system. In order to investigate thresholds and thus qualitative changes in multilingual systems with the help of a dynamic systems approach we will have to start from the ideal language acquisition curve and will first introduce language maintenance as a relevant parameter identified in multilingual systems development.

In Figure 20 language maintenance is mapped onto language systems development to illustrate the increase in LME resulting from an unspecified language acquisition process. The intermediate restructuring processes involved in acquisition are not shown in this idealisation.

Assuming that LME in monolingual language development tends to be sufficient it can in effect be ignored as a relevant parameter. As soon as we look at multilingual systems, required LME grows exponentially to the

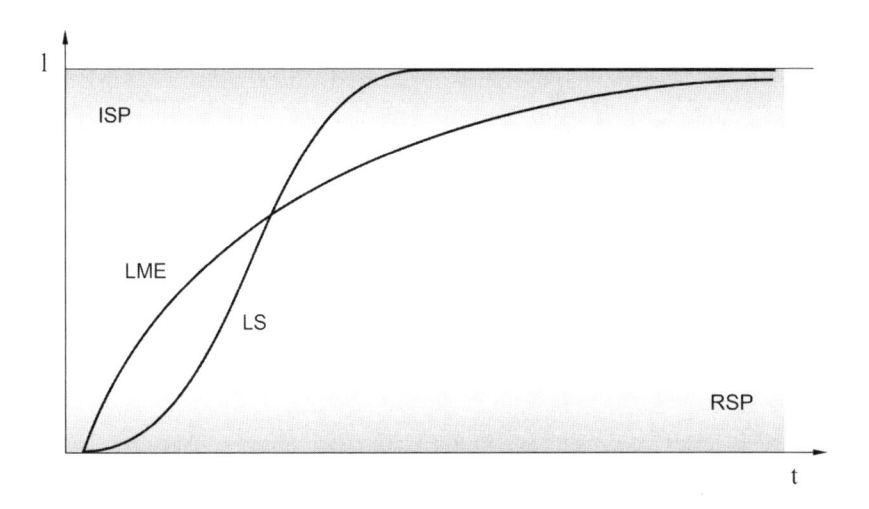

Figure 20 Ideal learning curve related to LME

LS = language system; LME = language maintenance effort; ISP = ideal native speaker proficiency; RSP = rudimentary speaker proficiency; t = time; l = language level

second order and soon exceeds the effort the average individual is prepared to put into the upkeep of her/his linguistic system. It therefore tends to impede the ideal language acquisition process as language growth tends to outstrip required LME, leading to gradual loss of the language system. This flattening of the growth curve is increased by the effect of transfer and interference, which not only affect performance but also the acquisition of the system itself, again leading to a general flattening of the learning curve.

Obviously the maintenance level normally does not vary excessively in an individual (differences in LME could, for instance, be motivated by a change of job or partner) and therefore has a stabilising effect on the learning curve as the language system does not collapse in the face of inadequate language maintenance, but sinks to a level compatible with the input. Inadequate language maintenance and interaction between the two or more language systems therefore leads to the development of a self-regulating homeostatic state, which can be seen as an approximative system of the target language resulting from an ideal learning curve (see Figure 21). The level at which a language system stabilises is not fixed and invariable, but must be seen as a self-organising system subject to constant variation. This result is likely to be obtained in the investigation of a

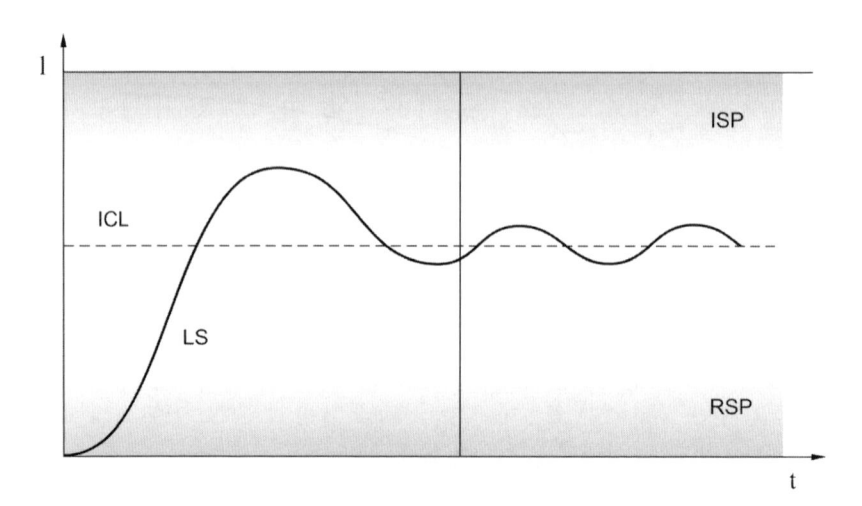

Figure 21 Approximative homeostatic systems

LS = language system; ISP = ideal native speaker proficiency level; RSP = rudimentary speaker proficiency level; ICL = input compatible level in terms of maintenance; t = time; l = language level

balanced bilingual speaker. The frequent alternative to this development is influenced by the factor of domain specificity or fossilisation as shown in the next figure.

Whilst Figure 21 illustrates partial achievement as specified by approximative systems theory, Figure 22 refers to a stabilising effect at the rudimentary level of a language system known as fossilisation. Domain specificity and fossilisation is shown by a relatively strong drop from native-like language competence to a restricted proficiency level with restricted vocabulary and syntax. These obvious disadvantages are outweighed by the fact that a relatively small drop in the LS level leads to a disproportionally larger drop in required language maintenance. Thus it is fairly easy for the multilingual speaker to maintain a proficiency level necessary for specific communicative functions, which is, however, far removed from native-speaker competence.

Yet we have noted that these multilingual impediments are counterbalanced by a number of advantages multilinguals have over monolinguals. One is the more pronounced development of monitor functions, which allow more effective use of the available language systems. If we therefore view language competence on the basis of proficiency results as described in Chapter 5, we will notice that bi- or multilinguals must be interpreted as

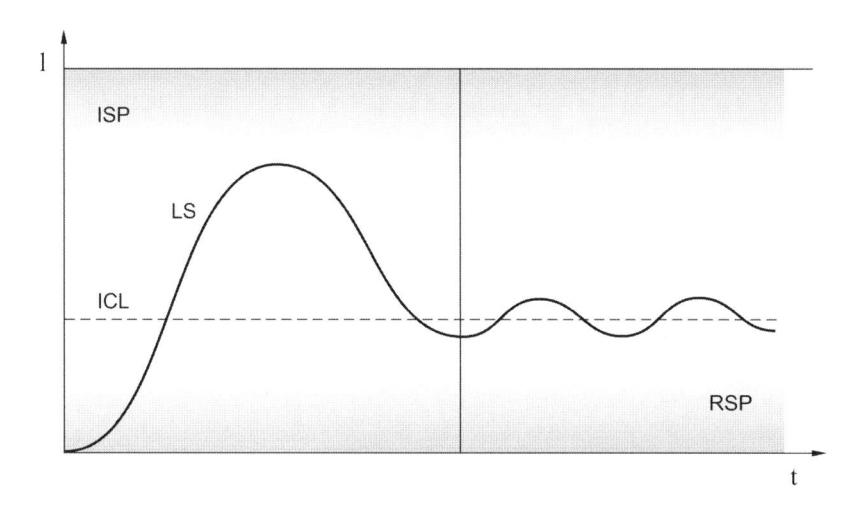

Figure 22 The effect of fossilisation on language systems development

LS = language system; ISP = ideal native speaker proficiency level; RSP = rudimentary speaker proficiency level; ICL = input compatible level in terms of maintenance; t = time; l = language level

overachievers, that is, performance results are likely to lie above the results to be expected on the basis of the analysis of the underlying linguistic systems. Figure 23 illustrates the plausible assumption that conscious monitoring processes can improve the language levels consistently attained.

The correlation between proficiency, that is measured multilingual language competence, and intelligence and/or general cognitive ability results from the significant function of the monitor in the multilingual speaker. Increased monitor functions required by bi- or multilinguals, as against the language acquisition process dealt with so far, are immediately related to the availability of cognitive functions. Less 'intelligent' speakers are thus less able to produce the necessary monitor functions for multilingual performance and are likely to underachieve. In our view this claim is compatible with the findings of modularity oriented research (see Chapter 8 on Christopher).

The final variation to be introduced in our graph results from the introduction of the common underlying proficiency hypothesis as described in Chapter 3 where we focus on metalinguistic abilities or awareness. As noted in Chapter 2, we would like to see the contact between the two

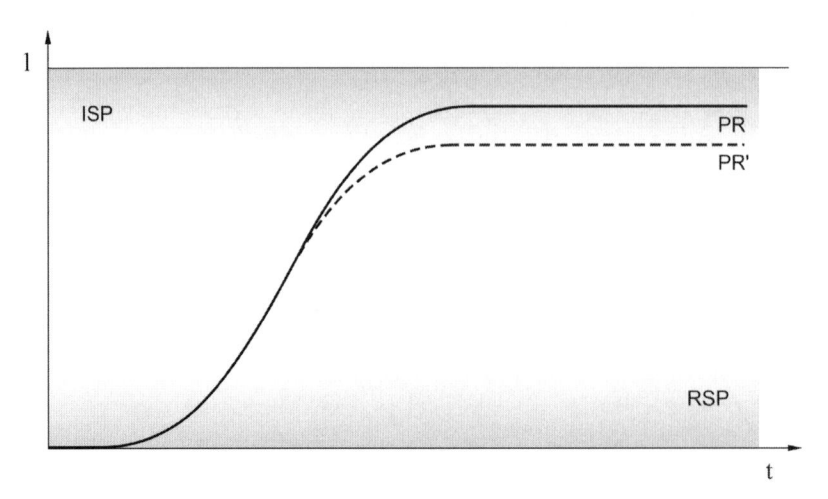

Figure 23 Language proficiency development relative to the learning process with and without monitoring function

PR = language development corrected by monitoring functions; PR' = projected language development without correcting influences; ISP = ideal native speaker proficiency level; RSP = rudimentary speaker proficiency level; t = time; l = language level

language systems not as a mere overlap of LS_1 and LS_2 but as a new quality which cannot be defined in terms of the parts of the two systems involved.

It is to be noted that metalinguistic abilities still lack the necessary operationalisation to be immediately empirically verifiable. We can nevertheless attempt to outline the effect metalinguistic abilities will have on the language learning curves used. It is important to realise that metalinguistic abilities, if a function of multilingual acquisition, obviously presuppose the existence of this phenomenon and are therefore difficult to observe in FLA and/or PLA, be it monolingual or multilingual. We will, however, expect these abilities to have a catalytic effect on further language learning processes, such as TLA undertaken by a multilingual speaker as shown in the next figure. Since it is not clear at this point of discussion whether (multi)language aptitude as described in the literature can be distinguished from metalinguistic abilities/awareness (see Chapter 5) we have decided to use MLA as acronym to include both concepts (see Figure 24). MLA constitutes one of the factors that can be interpreted as having a catalytic effect on the LS_3 or TLA. This can best be represented in terms of relative growth.

As can be seen from the last graph the MLA-hypothesis predicts an improved learning curve for TLA. Even if we were to assume that there

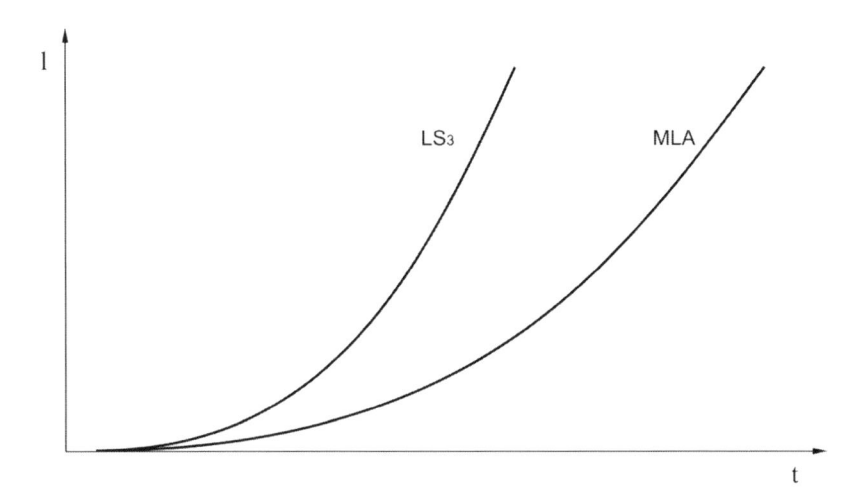

Figure 24 The catalytic effect of MLA on the development of LS_3 or TLA

MLA = (multi)language aptitude/metalinguistic abilities; LS_3 = third language system; t = time; l = language level

were sufficient evidence of the innateness of language aptitude – if this is not interpreted as to mean that some human beings have a language aptitude gene – then the innateness hypothesis says nothing about the nature of the property. In DMM we argue that aptitude is an emergent property of multilingual systems which is not directly predictable from genetic material or experience considered separately. Thus MLA is seen as acquired and not innate, although we might be able to agree on the existence of a number of dispositions facilitating the formation of skills and abilities which are found in the multilingual speaker in contrast to the monolingual.

Types of Multilingualism According to DMM

Now we would like to suggest a classification of multilingual types according to DMM based on the concepts described. As already stated in our definition of multilingualism in Chapter 5 we see bilingual systems as variants of multilingual systems but not equated with multilingual systems since multilingualism ranges from monolingual acquisition, that is the learning of an L2 by a native speaker, to balanced bilingualism or even ambilingualism and to the command of three or more language systems to point out a few stages. According to Cenoz (2000: 40–41) we encounter at least four possible acquisition orders in multilingual acquisition. The three language systems can be acquired consecutively (LS_1-LS_2-LS_3). Or the third language system can be acquired after the simultaneous acquisition of the first two referred to as primary learner multilingualism below (LS_1/LS_2-LS_3) or the first language system can already be acquired before the simultaneous acquisition of two other languages (LS_1-LS_2/LS_3). Another possibility involves the simultaneous contact with three language systems ($LS_1/LS_2/LS_3$).

As has been noted, bilingualism is frequently still assumed to be synonymous with ambilingualism (i.e. native-like competence in both languages), although researchers in bilingualism have identified a large number of forms depending on whether the respective definition concentrates on age of acquisition, order and sequence of acquisition, mental organisation, function, level of proficiency or/and social aspects involved (see Hoffmann, 1991: 18–27). In the description of bilingual types dichotomies play an important albeit problematic role since they are rarely mutually exclusive as already pointed out by Mackey (1969: 5). We shall here attempt to supply a systems-theoretic interpretation of a selection of the various forms of bilingualism and multilingualism and for our purposes we will focus on the balance between the systems of the multilingual speaker.

Balanced bilingualism

Let us first look at ambilingualism, with assumed native-like competence in both languages. It is one of the major contentions of this approach that ambilingualism in this sense of the term is most unusual as CLIN will lead to a distortion of performance, even if we assume that the speaker's language systems are both fully developed (due to increased communicative needs resulting from living in a multilingual society, bilingual community or professional demands). Obviously this effect can be compensated by increased monitoring although this may lead to different performance patterns, that is hypercorrectness in written form combined with an increased number of slips in speech.

In forms of balanced bilingualism we suggest to distinguish between ambilingual balanced bilingualism and non-ambilingual balanced bilingualism which in the literature, in Hoffmann (1991) for instance, are often referred to as ambilingualism and balanced bilingualism. Since the balance between the two systems in contact is crucial in both concepts, that is both forms represent balanced forms of bilingualism, we find the current use of terms somehow misleading.

Ambilingual balanced bilingualism (see Figure 25) is the simplest form

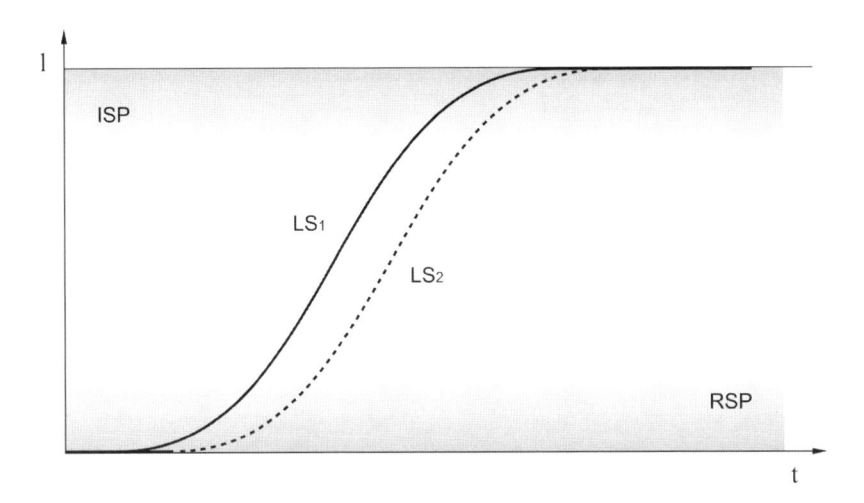

Figure 25 Ambilingual balanced bilingualism

LS$_1$ = first language system; LS$_2$ = second language system; ISP = ideal native speaker proficiency; RSP = rudimentary speaker proficiency; t = time; l = language level

of stable multilingualism. Note that it is assumed that both language systems are fully developed to an ideal native speaker proficiency level and that it is presented in non-simultaneous form to assist the graphic representation.

According to DMM non-ambilingual balanced bilingualism (see Figure 26) must be considered the more likely form of bilingualism where both systems are not considered developed to ideal speaker proficiency level. In non-ambilingual balanced bilingualism two language systems are equally developed but below a native speaker proficiency level. This also means to include rudimentary proficiency in both languages as described in immigrant children.

Balanced bilingual proficiency is likely to appear less developed and less homogeneous compared with monolingual competence. We are also less likely to encounter transfer phenomena (as these can be taken to depend on the dominance of one of the two language system), whilst interference effects will be less frequent. On the whole, however, we can assume ambilingualism to be a relatively rare phenomenon. Non-ambilingual balanced bilingualism can be assumed to be the more frequent form, although the fact that the determination of either ambilingualism or

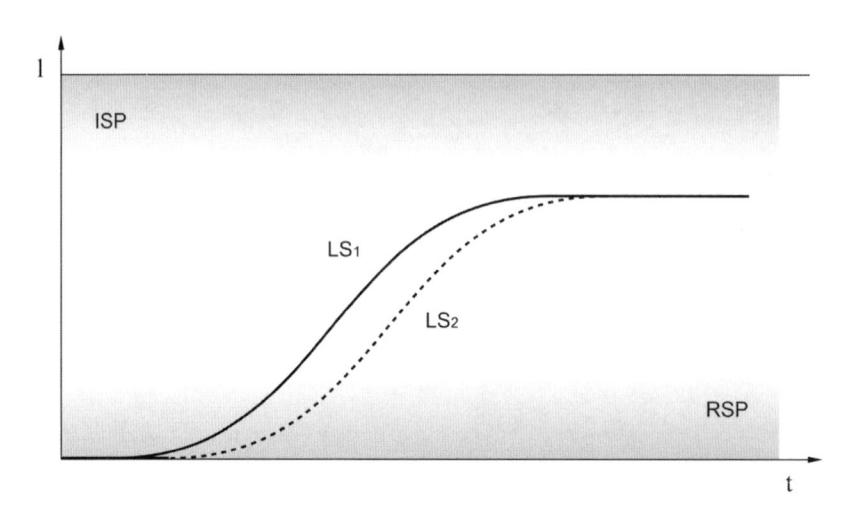

Figure 26 Non-ambilingual balanced bilingualism

LS$_1$ = first language system; LS$_2$ = second language system; ISP = ideal native speaker proficiency; RSP = rudimentary speaker proficiency; t = time; l = language level

balanced bilingualism has to be based on performance measures might make the distinction between the two types more difficult.

Unbalanced or asymmetrical bilingualism

On the whole we can expect to encounter dominant bilingualism which itself can be divided into transitional bilingualism and stable dominant bilingualism, far more frequently than balanced forms of bilingualism.

Figure 27 shows an extreme form of transitional bilingualism, where one language system is gradually replaced by another to result in a long-term reversion to monolingualism. The functions of LS$_1$ and the proficiency level of LS$_1$ are thus replaced by that of LS$_2$. In the case of transitional bilingualism we can observe a gradual transition from LS$_1$ to LS$_2$ as the dominant language system. This development is to be found in immigrant individuals rather than communities, that is individuals who due to outward circumstances lose their native speech communities and have to adapt to new communicative needs (i.e. is both perceived and objective) as specified later.

In the first phase of transitional bilingualism we will observe transfer phenomena occurring in the developing LS$_2$ whilst interference will be low

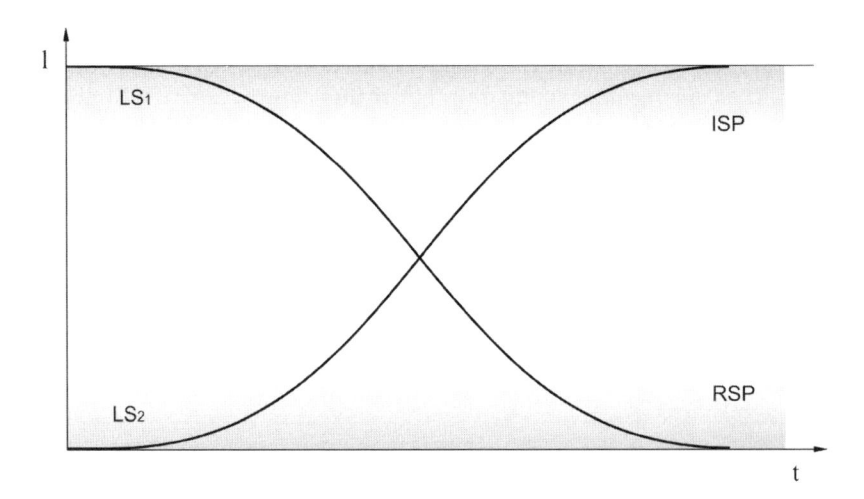

A Dynamic Model of Multilingualism Analysed 121

Figure 27 Transitional bilingualism

LS_1 = first language system; LS_2 = second language system; ISP = ideal native speaker proficiency; RSP = rudimentary speaker proficiency; t = time; l = language level

as we observe in normal SLA, in the second phase this process will be reversed as the LS_2 is by then sufficiently stable and the LS_1 is also undergoing a rapid process of erosion and has lost its originally dominant position. Here we observe a kind of backlash effect, such as phonetic transfer occurring in the LS_1, for instance the (bilingual) native speaker speaking her/his L1 with a foreign accent (see Flege, 1998).

Transitional bilingualism can be considered the most obvious proof of the validity of the maintenance effort hypothesis and the natural monolingualism hypothesis as in this instance LME can be seen as insufficient to maintain more than one full language system. As pointed out in Chapter 6, this problem is also seen by Seliger (1996: 617):

> When two languages come into contact within the same psycholinguistic environment, the speaker is forced to contend with the problem of the possible duplication of rules and functions in the two languages and the need to simplify this cognitive burden in some way.

There is, however, another possible development of the competing linguistic systems leading to the growth of a specialised language system, that is the development of a partial system, frequently referred to as freezing or domain specificity of LS$_2$ development.

This effect may be attributed to the fact that the effort put into LS$_2$ development, that is language acquisition effort (henceforth LAE), will at one stage begin to exceed overall available LME and result in backsliding (see Selinker, 1974: 36) leading to a reduction of systems-specific language proficiency in LS$_2$ (see below). Backsliding will continue to take place until LME is sufficient to maintain a steady state (see Figure 28). LS$_1$ is not fully replaced by LS$_2$, but the two systems coexist in a modified form, with LS$_2$ as dominant (primary) language LS$_P$ and LS$_1$ taking the role of subordinate (secondary) language LS$_S$.

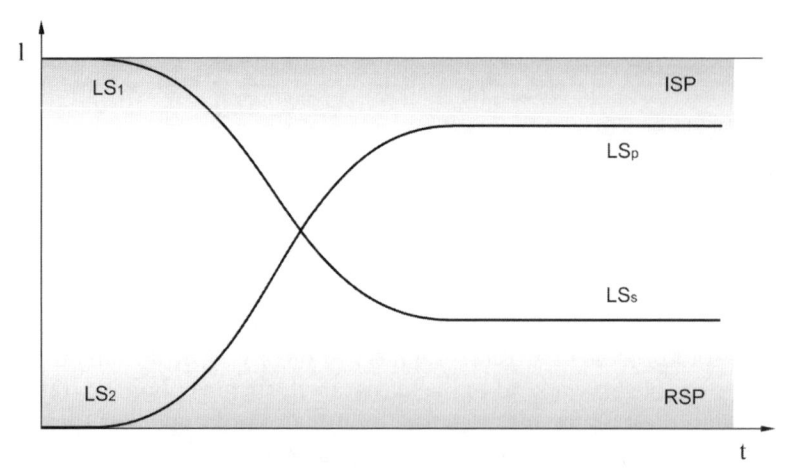

Figure 28 Stable dominant bilingualism

LS$_1$ = first language system; LS$_2$ = second language system; LS$_P$ = primary language system; LS$_S$ = secondary language system; ISP = ideal native speaker proficiency; RSP = rudimentary speaker proficiency; t = time; l = language level

It is assumed that the stabilising effect of LME is greater in a utility system than in learner pidgin. In essence, however, both utility systems and learner languages are interlanguages as described in Chapters 2 and 3, that is fossilisiation in a learner language makes it effectively indistinguishable from a utility or pidgin system (see McLaughlin 1987: 109–132). Whilst the utility system is no longer oriented towards a target language system, learner pidgin is still perceived as (communicatively) deficient by the speaker and her/his linguistic environment and is therefore more likely to fluctuate, as the multilingual speaker attempts to increase her/his general language effort to overcome these perceived deficiencies.

There are obviously other partial forms of multilingualism which are frequently cited but little researched. Partial achievement can be observed, when one of the linguistic systems has not reached a level sufficient to attain a steady state, that is a level of communicative efficiency allowing the system to attain a functional significance in daily communication, to allow the stabilising influence of the linguistic environment to take effect.

So-called passive bilingualism, in which the speaker has only passive/receptive command of one linguistic system and does not use LS_2 for active communicative purposes her/himself, can be taken as an instance of partial competence. It is to be noted that the difference between this form of partial competence and the competence of monolingual speakers involved in SLA is not categorical, the only significant difference lying in the fact that it is assumed that the LS_2 of monolingual speakers involved in learning a second language is still undergoing a growth process, that is, our expectations concerning the development of LS_2 are different. Our definition of partial achievement can be taken to imply that we do not expect the language systems involved to undergo great changes. The distinction between stable partial achievement and SLA may indeed be defined in relation to LME as will be explained in more detail below.

Multilingualism

In a multilingual context the number of languages and their acquisition order increase the complexity already encountered in bilingual systems. This picture is furthermore complicated if we assume that the acquisition of a particular language system might be interrupted and taken up later on again (Cenoz, 2000: 41).

The following exemplary developments of a multilingual system are supposed to provide an insight into our view of multilingual acquisition in DMM. Both figures refer to the multilingual development of the same speaker but whereas Figure 29a illustrates the development of the three language systems in contact from the moment when the multilingual learner gets in touch with the two foreign languages, Figure 29b also concentrates on the prior language knowledge of the speaker and how this knowledge effects further language acquisition. Thus Figure 29a can be detected as an extract of Figure 29b.

In Figure 29a the primary language system is considered constant and dominant at the level of ideal native speaker proficiency whereas the relationship between the secondary system and the tertiary language system

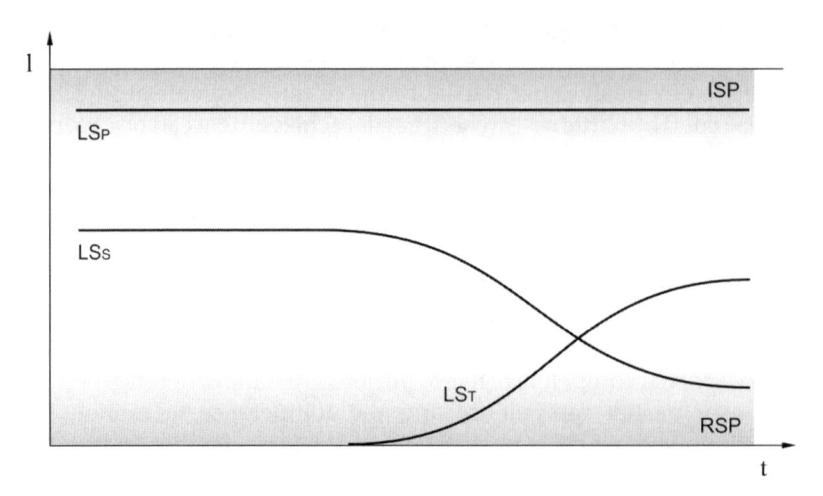

Figure 29a Learner multilingualism: acquisition phase
LS_P = primary language system; LS_s = secondary language system; LS_T = tertiary language system; t = time; l = language level

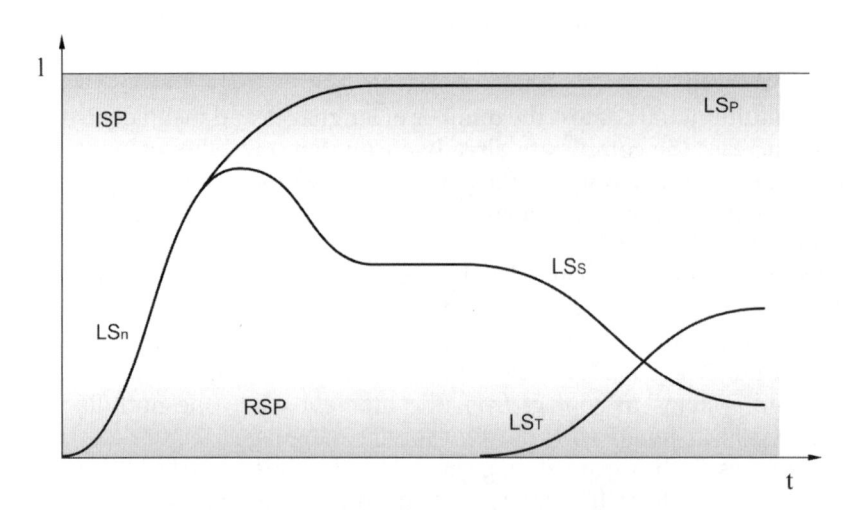

Figure 29b Learner multilingualism: overall development
LS_n = prior language system(s); LS_P = primary language system; LS_s = secondary language system; LS_T = tertiary language system; ISP = ideal native speaker proficiency; RSP = rudimentary speaker proficiency; t = time; l = language level

can be described as transitional. The focus in Figure 29b is on the effect of TLA on preexisting language systems.

In the standard form of learner multilingualism LS_n (n to signify that within DMM the attribution of LS_1 is arbitrary) is normally given and dominant, whilst the learner system LS_s is still incipient and undergoing development. The types of learner multilingualism are obviously dependent on the acquisition of the first two languages. There are several possibilities of distinguishing the types of bilingualism involved in third language learning. We might therefore distinguish between full bilingualism, where LS_p and LS_s are seen as full language systems, and learner bilingualism where LS_s is evidently a learner system. Obviously there are many possible means of distinguishing types of bilingualism, as for example according to whether bilingualism is simultaneous or successive (see transitional bilingualism above), or in other words whether we are confronted with primary bilingualism (LS_p acquired with LS_s) or secondary bilingualism (LS_s acquired after LS_p). Primary learner multilingualism, that is a third language is learnt after primary bilingualism, can be described as the less frequent albeit increasing form of learner multilingualism. Secondary learner multilingualism refers to the acquisition of two languages simultaneously after LS_p has been acquired.

It will have been noted that DMM does not in essence distinguish between multilingual systems containing two languages and multilingual systems containing more than two languages. It is assumed that the same basic principles apply to SLA multilingualism as to TLA multilingualism. But as indicated before specific differences are nevertheless to be expected.

How Factors Relate in DMM

As was pointed out at the end of Chapter 6, our analysis so far has tended to view multilingualism as a homogeneous and unitary phenomenon, which researchers on multilingualism know it not to be. In fact, Grosjean (1992) claims that bilingualism and thus multilingualism is not so much a clearly identifiable state, as it is a type of progress along a continuum. Yet, Grosjean assumes that there is at least one identifiable constant in this process, and that is what he terms 'a necessary level of communicative competence.' He does, however, admit that this constant is largely dependent on the communicative needs of the speaker as defined by her/his environment. The idea of a constant level of communicative competence, however, might appear to be confusing, if not erroneous (see Hoffmann, 1991: 151).

Firstly, we have identified at least one factor leading to a variation in the level of communicative competence and that is the communicative needs of the speaker. Another critical factor introduced in the previous chapter is LME which can admittedly be related to the communicative needs of the speaker, although these would have to be defined more specifically as the perceived needs of the speaker and not objectively assessed needs. Obviously, language learning cannot be interpreted in homeostatic terms, that is the replacement of knowledge of the LS_p by that of the LS_s, but has to be interpreted as a growth in communicative proficiency, limited by factors as outlined above.

We may therefore assume the level of communicative proficiency to be variable rather than constant and dependent on LME as a function of the perceived communicative needs of the speaker. This will correlate with, though not be identical with, the social status of the speaker (i.e. the kind of job held by the speaker, the communicative environment accompanying a social position, such as degree of isolation, public relations, etc.) and the sociolinguistic parameters of the speaking community.

We must further note that it is erroneous to identify underlying language systems with observed performance. Although this statement obviously also applies to (monolingual) native-speaker utterances, the problem is accentuated in multilingual speakers for the reasons given. We have already noted that tested or observed performance is likely to lie below the level of the existing language systems as language contact will lead to a certain number of language processing errors. How this crosslinguistic deviation can be isolated from errors to be taken as evidence of a deficient language system represents a complex methodological problem.

The systems model obviously implies certain relations between the factors mentioned, which we shall now attempt to isolate and treat individually. As a preliminary step individual factors will be related to others by describing them as 'functions of,' which allows various formal interpretations of these relations in part at least only to be determined on the basis of empirical evidence.

The bilingual system as a function of LS_1 and LS_2

The bilingual language system can be taken as a function of LS_1 and LS_2. Note that it is not simply the result of adding LS_1 and LS_2 as we know that bilingual performance is also determined by CLIN and MLA, for example. We must also distinguish between theoretical overall communicative proficiency (non-system specific) – in which the multilingual's communicative competence is bound to be superior to that of the monolingual as the assessment of overall communicative efficiency will always depend on the communicative environment of the speaker and does therefore not provide a sufficiently independent criterion – and systems-specific communicative efficiency which constitutes the only possible valid internal criterion.

Communicative efficiency must therefore always be assessed in terms of the existing language systems employed by the bilingual. We are also at this stage unable to determine the degree of disturbance caused by crosslinguistic interference as part of CLIN or the catalytic effects of MLA. We must also draw attention to the fact that we measure proficiency on the basis of measured bilingual performance observed.

$$LS_1 + LS_2 \cong BLS$$

$$BP \cong BC + CLIN + MLA$$

where LS_1 = first language system; LS_2 = second language system; BLS = bilingual language system; BP = bilingual proficiency; BC = bilingual competence; CLIN = crosslinguistic interaction; MLA = (multi)language aptitude/metalinguistic abilities

Multilingual proficiency as projection

As argued above, DMM requires an essential clarification of what is to be understood by the command of a language or language system. Within DMM it is obvious that the simple competence/performance distinction introduced by Chomsky is not sufficient. As performance is essentially intended to cover unsystematic deviation from language competence, learners and bilinguals are characterised by systematic deviation from the performance expected on the basis of an LS. Whilst learner deviation can be dealt with in terms of the 'independent grammar assumption' (see wild grammar discussion in Chapter 4) the systematic deviation in relation to the respective speaker systems cannot simply be attributed to lack of competence.

The simplest concept is that of full attainment, which means, for example, that the bilingual speaker has full command of (native-speaker-like) language systems LS_1 and LS_2 and any systematic deviation from the expected language norms is to be attributed to the interaction between the two systems. Unsystematic deviation can be attributed to performance or differing monitor function responsible for language management, that is, the skills which a multilingual develops in order to integrate differing language resources into her/his repertoire and to keep them apart at the same time in a communicative act. The communicative requirements have thus to be balanced with the speaker's linguistic resources.

Partial attainment can affect either only one LS or both language systems at the same time. Deviations measured in performance can, however, be attributable to CLIN as much as to the specific level of partial attainment and are also a function of systems stability. If we furthermore assume that the monitor fulfils a compensatory function, we can expect better performance results in more prepared situations. There is, however, no immediate correlation between observed performance and specific language attainment. Within a multilingual system of more than two languages language management will play an increasingly important role, making the relationship between observed performance and underlying specific language competence an even more complex one.

If we introduce a communicative efficiency measure of which both proficiency and competence must be considered subsets, the measure of multilingual proficiency in a monolingual society will always be in terms of how well it measures up to assumed monolingual competence. In a multilingual community on the other hand the speaker's communicative efficiency will be measured in terms of how well s/he can communicate in either or both languages (command of LS_1 and LS_2). Using a bilingual measure, however, the monolingual speaker must be considered merely half as efficient as the bilingual speaker. We will thus need a cumulative measure of communicative efficiency as suggested in Figure 30.

In this figure we illustrate the fact that multilingual proficiency should be measured in cumulative rather than in monolingual terms. The dotted

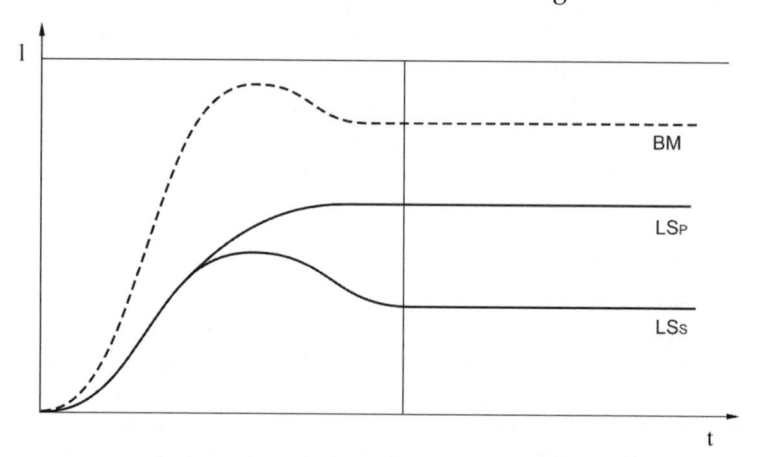

Figure 30 Cumulative measure of multilingual proficiency

LS_P = primary language system; LS_s = secondary language system; BM = bilingual measure of language proficiency; l = language level; t = time

line indicates a supposed level of multilingual proficiency based on two languages according to the concept of bilingual measure of language proficiency.

The M-factor or properties specific to multilingual systems

Finally we assume that well-developed multilingual language systems will lead to a number of factors distinguishing them from monolingual systems, that is, the multilingual learner develops new skills, such as language learning skills, language management skills and language maintenance skills which are linked to a change in quality to be expected in the language learning process. All these skills can be seen as contributing to what has been termed metalinguistic awareness. This aspect which can be seen as enhanced in multilinguals has been identified as one of the advantages multilinguals develop in contrast to monolinguals. In Chapter 5 we have already identified an enhanced multilingual monitor (henceforth EMM), where we can presume that the development of the monitor is directly proportional to the number of competing language systems available to the speaker and the frequency of (alternate) use of the systems.

In DMM it is furthermore assumed that in contrast to monolingual systems the development of the EMM is dependent on general cognitive functions. The development of an EMM is therefore necessarily restricted by the general cognitive ability of the individual speaker. This assumption is indeed supported by the findings of Smith and Tsimpli (1995), as Christopher, the subject of the study, being exceptionally linguistically gifted although cognitively impaired, has developed a surprising number of (albeit rudimentary) language systems on the one hand, but has substantial problems managing (or monitoring) the language systems at his disposal (see also Chapters 4 and 8). The lack of EMM, for instance, explains the difference between his translation skills on the one hand and his conversation skills on the other.

$$EMM = f(GCA)$$

where: f = function; EMM = enhanced multilingual monitor; GCA general cognitive ability. Note that GCA is not assumed to be the same as IQ, although results in tests of intelligence are taken to correlate with general cognitive abilities.

Finally, as we have already suggested above, EMM is part of the M-factor through which, for instance, the bilingual speaker is advantaged when acquiring LS_3. Our systems-theoretic approach thus predicts the existence of an M-factor, a dispositional effect which will have a priming or catalytic effect in TLA. The M-factor as an emergent property can be specified as a function of the interaction between more than one language system. It is not necessarily relevant how these language systems develop, but may be dependent upon the number of language systems involved.

M-factor = f (of n)

where: M-factor = multilingualism factor; f = function of; n = number of LS in a multilingual system.

Within a bilingual system this factor obviously consists of a dispositional measure which will primarily be observable on the acquisition of a further language. It implies that we expect a difference in the development of communicative efficiency exhibited by the monolingual and the bilingual speaker in TLA. Furthermore the M-factor can be interpreted as having the same effect on the acquisition of a third language as its component MLA (see Figure 24). This effect could best be represented in terms of relative growth which marks the difference between the expected and the accelerated development of the third language system.

The multilingualism factor expresses an essential difference between multilingual and monolingual speakers. We must assume that the multilingual system:

(1) contains components the monolingual system lacks and
(2) that even those components the multilingual system shares with the monolingual system have a different significance within the system.

One of the most salient differences concerns the arbitrariness of signs which only becomes apparent if the speaker comes in contact with a second language system.

Our formal representation of the multilingual system will thus include a further factor M, apart from the factors corresponding to the respective language systems of the multilingual speaker and a factor representing the effect of the interrelation between the language systems. The following formula crudely expresses a possible relationship between competence, performance and proficiency in a multilingual speaker:

$$LS_1 + LS_2 + LS_n + CLIN + M \cong MP < p$$

If we define competence in terms of the command of a specific language, we can derive the following formula from the above:

$$C_1 + C_2 + C_n + CLIN + M \cong MP < p$$

where: $LS_1/LS_2/LS_n$ = language system; $C_1/C_2/C_n$ = competence in a particular language; CLIN = crosslinguistic interaction; M = M-factor; MP = multilingual proficiency; p = performance.

Thus the M-factor, as an emergent property of a multilingual system, in the given formula refers to proficiency skills as developed in the multilingual speakers. These skills show several characteristics clearly distinguishing the monolingual from the multilingual speaker and are taken to include skills in language learning, language management and language maintenance. This metasystem developed in multilinguals is the result of its relation to a bilingual norm, that is TLA relates to a system containing two languages whereas SLA relates to a monolingual norm.

General language effort as a function of language acquisition effort and language maintenance effort

The degree of multilingual proficiency observed in the multilingual is defined as an indirect function of general language effort. GLE itself is conceived of the sum of LME and LAE as the effort put into language development already mentioned above.

$$GLE \cong LAE + LME$$

where: GLE = general language effort; LAE = language acquisition effort; LME = language maintenance effort.

Although it is generally assumed that LAE is not immediately relevant to stable bilingual systems or multilingual systems, we cannot ignore LAE for two reasons:

(1) LAE is obviously a necessary presupposition of the development of monolingual or multilingual systems and
(2) we can relate LAE to LME; that is, assuming GLE is constant, we can claim that LAE has an inversely proportional function to LME.

As the LS grows due to increased LAE, more and more of the GLE is consumed by the LME required to maintain the existing LS. The only exception to this rule appears to be the development of partial systems, where it is assumed that reduced communicative demands lead to a reduction in LME, bearing in mind that LME exhibits second order growth, when related to growth of multilingual proficiency. Thus the distinction between stable partial achievement and SLA may be defined in terms of a function of the respective proportions of LME and LAE in GLE, that is as long as LAE exceeds LME, we may speak of SLA, whilst if LME greatly exceeds LAE or GLE consists exclusively of LME we may speak of partial achievement.

Figure 31 shows the interdependence of LAE and LME with GLE as a

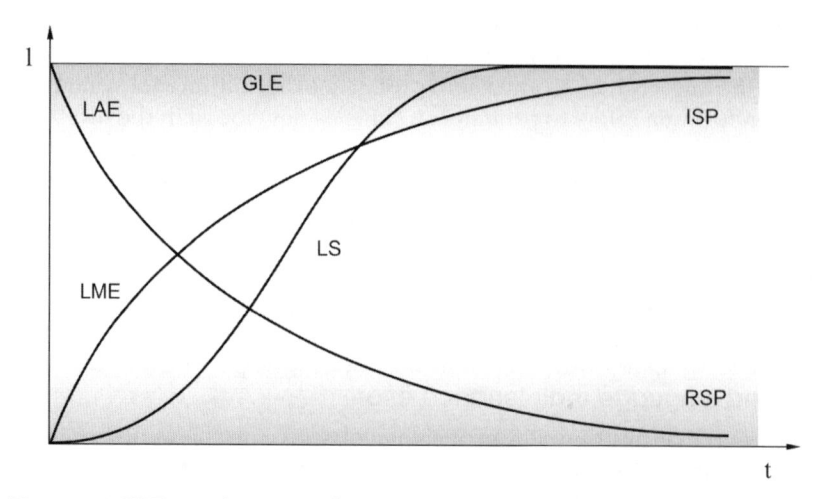

Figure 31 GLE as a function of LAE and LME

GLE = general language effort; LAE = language acquisition effort; LME = language maintenance effort; ISP = ideal native speaker proficiency; RSP = rudimentary speaker proficiency; LS = language system; t = time; l = language level

constant (see also Figure 18). Although we assume that bilingualism is the simplest form of multilingualism, essential changes take place in the learner or speaker of a language, as soon as the number of language involved exceeds two. This can be attributed to the fact that acquiring two languages (sequentially rather than simultaneously) leads to the development of specific metaskills, concerning the acquisition of language systems as a whole that will certainly affect the language acquisition process. On the other hand three or more languages obviously constitute a more comprehensive or heavy language load for the respective speaker, which has an influence on language stability and LME, the effort required to maintain a working multilingual system. Obviously individual dispositions will have a definite effect on the capacity of an individual to acquire and maintain a substantial number of languages. Note that the formalisation expresses a complex dependency relationship by which communicative needs are to be seen to determine the multilingual system and consequently show their influence on its stability.

Systems stability

There are obviously a number of factors influencing or determining systems stability which are themselves dependent on whether the language system is in transition or relatively stable. Stability can be expressed as a function of the rate of effective change on the one hand, but also the rate of expected change on the other hand. This means that the speaker's expectations concerning future language requirements can be seen as having an effect on the stability of the multilingual system. Furthermore the functional relation of multilingual proficiency as described above is necessarily incomplete as the degree of CLIN must be seen in rather more indirect terms as a function of language systems stability.

We can also define relative systems stability as a function of the language systems involved. Within DMM systems stability is necessarily not only a function of the system in isolation but the competing systems as such. As predicted in DMM, we can define interference as a function of the proximity of the language systems involved, that is, we predict interference to be more likely to occur in balanced systems as stated in Chapter 3.

We first specify language systems stability as a function of the command of a language. We will subsequently define relative systems stability as a function of the proximity of the two or more language systems. The closer the levels of both systems are to each other, the greater the degree of interference to be expected as systems integrity is threatened by the degree of proximity. Transfer on the other hand is defined as a function of the distance between the two (or more) systems.

As predicted by DMM, transfer will therefore be reduced by an increased proximity of the systems. Please note that in contrast to the assumption of equicomplexity the assumption of equidistance (both explicated in Chapter 6) is obviously not to be taken into consideration in this discussion of distance typology.

Positive and negative growth in DMM

Variation in DMM can be viewed in two stages: firstly variation within a primarily stable system and secondly variation within a dynamic system. Within a fully dynamic system the relations soon attain a degree of complexity that becomes difficult to model.

First we should view language variation with certain parameters taken as constant. As outlined in DMM the prime criterion determining language growth is taken to be GLE comprising LAE and LME. If GLE is taken to be constant, we can observe two predictable effects: as LAE leads to positive growth in the language system being acquired, so the effort required for language maintenance increases, effectively reducing the amount of LAE available to contribute to further language growth. There is therefore an inversely proportional relationship between LAE and LME. This principle will contribute toward the explanation of accelerated language acquisition in childhood as there is no pre-existing system requiring a certain amount of GLE for the maintenance of an existing language system as shown in our graph modelling an example of overall development of learner multilingualism (Figure 29b).

The logical consequence of what has been said above is that the acquisition of a second (or third) language will place new demands on the GLE of the speaker. If the speaker is unable to increase the amount of GLE available, this will mean that the speaker has to redeploy GLE to accommodate the demands of the new system. A multilingual system including an LS_1 and a learner system will consist of an LME for LS_1, an LAE for LS_2 and a smaller LME for LS_2. If the speaker's overall GLE is insufficient to support the coexistence of two or more language systems, we will observe the phenomenon of transitional bilingualism, that is, one language system being replaced by another language system following a period of transition. These transition stages indicate a qualitative change in the state of the overall system, which can also be interpreted as a threshold.

In Figure 32 the phenomenon of transitional bilingualism is presented in terms of stages, that is stable stages interrupted by stages of transition. The stability of the system is assumed to be influenced by interference and transfer which are themselves dependent on the (psychological) distance between the language systems involved.

As gradual language loss is retarded negative growth (represented as a curve asymptotically approaching the x-axis) we may assume that language maintenance is not directly, but indirectly related to the level of

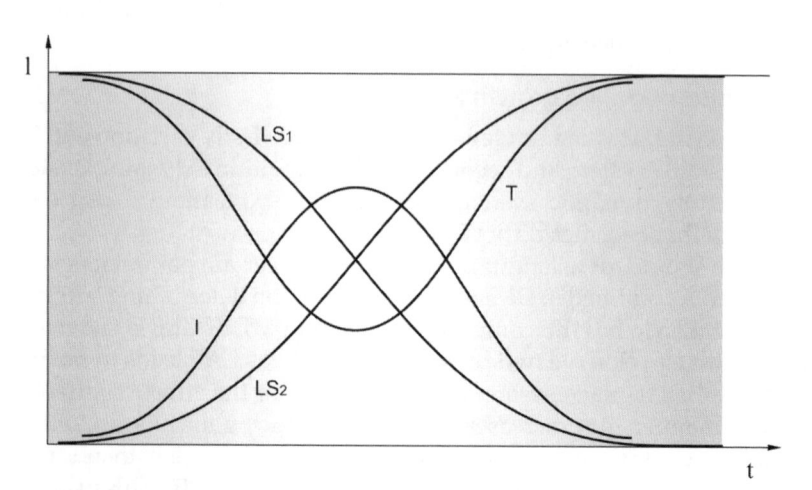

Figure 32 Transition period

LS_1 = first language system; LS_2 = second language system; I = interference; T = transfer; t = time; l = language level

the language system, that is, first order growth of the language system leads to second order growth in language maintenance. This relationship would explain why a complete loss of a language system (which was once part of competence) is unusual (except in clinical contexts). We must obviously also assume that gradual language loss is a delayed effect of gradual loss of language use.

The speaker might of course counter the greater demands of concurrent language systems by increasing her/his GLE to cover the effort required by the simultaneous maintenance of two or more language systems. Within a world of finite resources of time and effort we must assume the existence of certain motivational factors determining the speakers decision to increase her/his GLE. According to DMM the key to the increase in GLE lies in his or her communicative needs determining positive and negative growth as explained in the next chapter.

Communicative needs

When developing a realistic model of the multilingual speaker, we must assume that the speaker is substantially influenced by her/his linguistic environment. One of the hypotheses of DMM is that the main factor determining the development of the multilingual system, GLE, is determined by the speakers perception of his/her communicative needs. That is, the greater the (multilingual) communicative needs perceived by the speaker, the greater the (general language) effort of the speaker to meet these requirements. Note that these needs are not reducible to the effective communicative needs at a specific point in time, but must always allow for the possibility of the anticipation of a different (world) state. Communicative needs refer to all situations in which language is used as a means to express oneself or as a tool for thought and thus also can refer to the learning of dead languages such as Latin or languages which are acquired without primarily having the need to use them for communication with other individuals.

We thus obtain two factors determining GLE: the effective communicative needs of a speaker A in a situation B (for which we shall suggest suitable measures) and, on the other hand, the perceived communicative needs of the speaker. Note that these two factors are necessary for the dynamic development of the multilingual system, as language acquisition is necessarily based on the anticipation of communicative needs not present at a particular point in time. Let us, however, first turn our attention to effective communicative needs.

Effective communicative needs

Effective communicative needs are determined by the recurrent communicative requirements of the speaker. In a multilingual environment it is determined by:

- the number of communicative exchanges with speakers of language L1 and speakers of language L2,
- the duration of the exchange and
- the intensity of the exchange, which can be determined as the closeness of the individual with whom the exchange has taken place.

In evaluating the communicative needs within a bilingual community, we will thus be able to obtain a factor indicating the proportional requirements of L1 and L2.

Perceived communicative needs

As already pointed out, perceived communicative needs are effective in introducing change into a language system or resistance to change in the face of pressure. In the latter instance we might observe that a speaker is resistant to the requirements of her/his communicative environment on the basis of an implicit decision not to require a specific degree of knowledge of L2 (see the case of Alberto as referred to by Schumann, 1978).

On the other hand perceived communicative needs also allow for the anticipation of counterfactual conditions or possible worlds in which the effective communicative requirements are other ones. A speaker learning a language will anticipate a situation in which her/his language requirements will have changed, that is, s/he will be required to communicate with speakers of an L2 to a greater extent than is the case at present. This prototypical situation can be termed the anticipation of communciative needs. The perception of communicative needs can, however, also be delayed, that is, the speaker will find her/himself in a situation where communicative needs are found to differ from those which s/he is accustomed to. This will probably also lead to adaptive behaviour which will be expressed in an increase of GLE or in a shift of LME from one language system to the other.

Thus GLE (the sum of LME and LAE) is not an immutable parameter as it is itself determined by various factors such as life style, social status, economic aims and perceived social and communicative needs. GLE can therefore be defined as a function of the perceived communicative needs (henceforth PCN) of the multilingual speaker, which themselves indirectly relate to the effective communicative needs of the speaker. The effective communicative needs of a speaker primarily determine maintenance levels and the perceived communicative needs allow for an individual interpretation of reality and anticipation of requirements in terms of subjective language planning, etc. (thus primarily determining LAE).

$$GLE = f(PCN)$$

$$PCN = f(CN)$$

where: f = function of; GLE = general language effort; PCN = perceived communicative needs; CN =effective communicative needs. Thus a multilingual system can indirectly be interpreted as a function of communicative needs.

One might easily be confused by the apparent coexistence of personal and social factors. This problem is best solved by assuming that the personal factors represent the internal environment of the speaker whilst the social factors point to the external language relevant environment of the speaker. In this sense the personal environment can be seen as embedded in the social environment, where for example perceived communicative needs on which language development may rest will relate to, but not necessarily correspond directly to the more objective (i.e. effective) communicative needs that can be derived from the speaker's environment. Apart from these two main factors there are of course a large number of personal ones determining the development of the language systems. To these we shall turn our attention to now.

Personal or psychosocial factors affecting multilingual proficiency

So far DMM has dealt with idealised or standardised growth models allowing for systematic variation in the development of language systems. It is assumed that multilingual language development in individual speakers complies with one of the prototypical forms outlined by DMM. In contrast to other models of multilingualism, DMM admits of systematic variation in language development and gradual language loss within a multilingual system but it must also attempt to explain individual deviation from prototypical variation or second order variation to be attributed to individual factors.

We have therefore attempted to construct a model of individual factors involved in developmental changes. It is important to note that this model has exemplary character and does not claim to specify all the actual factors involved in the process. The factors determining language acquisition progress or more precisely, the rate of change in positive or negative language growth, can, according to Figure 33, be subdivided into motivational factors, perceptional factors and anxiety. The graphs are intended to

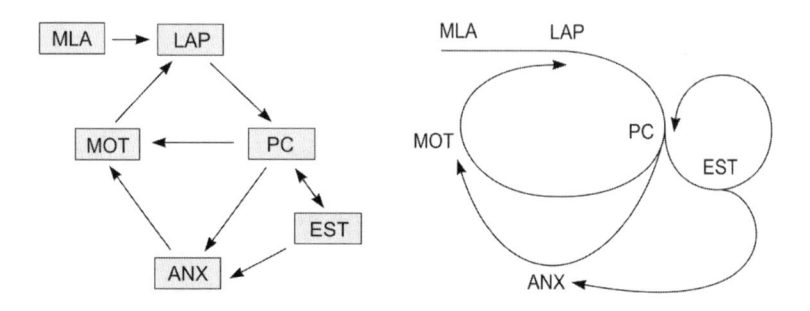

Figure 33 Some individual factors involved in the development of a multilingual system

MLA = (multi)language aptitude/metalinguistic abilities; LAP = language acquisition progress; MOT = motivation; ANX = anxiety; PC = perceived language competence; EST = self-esteem

illustrate the type of complex interdependences we are to expect between the factors involved in language acquisition processes. The relationship between the exemplary factors determines the rate and direction of the development of the new language system. The presentation on the right is intended to illustrate the autodynamic behaviour of the system.

The factors can be explained as follows:

- (multi)language aptitude/ metalinguistic abilities (MLA)

As already discussed before (see Chapter 5 and 7) this factor is taken to influence the development of the language system. It forms part of the M-factor as defined in DMM.

- language acquisition progress (LAP)

Language acquisition progress can be positive or negative, in the latter case expressing the speed of individual language loss. It is primarily an indicator of the rate of change of a language system.

- motivation (MOT)

Motivational factors are seen as the prime factors in determining general language effort. We can thus assume that individual motivation will show its effects on the amount of effort put into the acquisition and maintenance of a specific language system and therefore on positive or negative growth (see Gardner *et al.* 1987; Dörneyei, 1998).

Motivation, whose neurobiological foundation has been discussed by Schumann (1997), will be determined by the speaker's PCN and the speaker's interpretation of in how far her/his language competence is able to meet the perceived communicative needs. It is important to note that as a psycholinguistic model DMM distinguishes between effective communicative needs and perceived communicative needs, and effective competence and perceived competence, that is degree of motivation available can be interpreted as resulting from the difference between the speaker's perceived competence (how good s/he thinks s/he is) and the desired goal (how good s/he thinks s/he ought to be). This dynamic concept of motivation (see also Gardner & MacIntyre, 1992; 1993), that is reciprocal determination between motivation and achievement, has been underlined by recent research (Ushioda, 1996).

- perceived language competence (PC)

The greater the perceived language competence in the language system to be developed, that is, the more the speaker is perceived to meet her/his communicative needs the smaller the effort s/he is going to put into language acquisition as explained before. Note that this does not necessarily imply a minimalist interpretation of communicative needs (i.e. according to a minimax principle known from game theory) as the speaker will (re-)evaluate her/his communicative needs accordingly. Perceived language competence is therefore not necessarily identical to language competence as measured or attributed by others.

- self-esteem (EST)

Self-esteem or self-confidence is a significant factor in determining the willingness of the individual to engage in communicative activity. It is dependent on the PC or high self-ratings of individual language proficiency as communicative failure within a language community is to be avoided. This factor is also closely related to language anxiety, that is, '[t]his factor describes an individual who is not anxious when using English, has prior experience in doing so, and is self-assured with respect to his or her own English proficiency' (Clément & Smythe & Gardner, 1980: 298).

- anxiety (ANX)

Anxiety is not merely a performance-related phenomenon (leading to communicative underachievement) but acquisition-related as anxiety will also reduce the toleration of communicative blunders, without affecting self-esteem (see also MacIntyre & Gardner, 1989; 1991a,b; Gardner & MacIntyre, 1994).

As the developments of multilingualism outlined above are essentially prototypical, the individual factors will necessarily have an effect on overall language development with either an accelerating or a retarding effect on language development (see Ellis, 1994: 467–527; Ehrman & Oxford, 1995). It is important to note that in the outlined prototypical developments of multilingual systems these factors are taken to be constant, that is, not to affect the course of the development as such. A significant change in one of these factors will affect the development of LS_1, LS_2, etc. resulting in very complex patterns of language development, explaining the wide variety of multilingual proficiency observed. For instance, van Geert (1994: 192–194) describes a threshold model of self-esteem where self-esteem going beyond a realistic and acceptable level is considered negative for learning and achievement.

Additionally it is important to note that these factors do not only show their influence on language growth but also depend on each other in more than merely unidirectional relationships. The feedback loops contained in the illustration are supposed to show that the system has the ability to determine its own conditions of growth as part of its autopoeitic behaviour. The growth and size of the input will determine the rate of growth.

It has already become clear how the complexity in the developmental changes of multilingual systems has increased in DMM with the introduction of the various influencing factors by paying special attention to their development and relationship to each other. The modelling of such complex systems poses theoretical problems which have to be addressed before we can move on to the issue of holism in DMM.

Mastering Complexity in DMM

The models outlined in DMM are idealisations in more than one sense. DMM patterns are ideal-typical in so far as they do not not try to model individual developments of multilingualism but rather suggest an abstract idea of how a particular model of multilingualism would develop if we ignored individual variables. This does not mean that the models are empirically irrelevant or the predictions incorrect, as the introduction of personal variables will then allow the prediction of individual multilingual development as a second stage of theory development. Obviously the prediction of individual multilingual development presupposes the existence of an abstract general model (see van Geert, 1994).

This abstract model by necessity works with idealisations which are best identified with the concept of ideal types originally introduced by Max Weber.

> [I]deal types cannot be understood as descriptions of historical or cultural reality. They are rather hypothetical and counterfactual constructions of possible realities in which the diffuse properties of real phenomena are synthesised on the basis of value relevancies. Weber notes that the empirical factors that make up the complex of elements that define a certain ideal type rarely, if ever, occur together to the same degree. Indeed, there may be instances in which some elements of the complex may be absent altogether [...]. However, these characterisations do not compromise the theoretical value of ideal types as concerning how phenomena would behave if they were conceptualised on the basis of certain values. (Oakes, 1994: 1639)

In a mathematical model these ideal types are obviously to be identified with attractors, an idea also adopted by Larsen-Freeman (1997: 146). The reason why we stress the counterfactual nature of these ideal types is based on the fact that at least the formulae provided can be (mis)interpreted as static algorithms of which Larsen-Freeman (1997: 149) states:

> A static algorithm cannot account for the continual, and neverending growth and complexification of a system that is initiated from the bottom up. It cannot account for the performance 'inconsistencies of competing dialects and registers', nor the 'improvisational metaphors of ordinary language usage' (Diller 1995:112). To do so, a dynamic model of performance is needed, which relates individual use to systemic change.

Furthermore DMM patterns are ideal-typical as they are necessarily simplifications of the development of multilingual systems even as an abstract model, some of which, the homogeneous multilingual proficiency assumption, the homogeneous growth assumption, the equidistance assumption and the equicomplexity assumption, have already been described in Chapter 6. Another crucial aspect is that the patterns outlined work on a simplified assumption of limited resources and the competition of the language systems for these resources. This assumption allowed us to predict the conditions of transitional multilingualism, gradual language loss, etc. and thus represent necessary and useful idealisations. The idea of GLE works on the assumption that the effort available (although variable) is to be divided up between the languages the multilingual speaker has to maintain. The same applies to LAE as a subcomponent of GLE.

What has been discussed as one of the characteristics of dynamic systems but has been included in the modelling so far only to a moderate extent, in the discussion of the M-Factor for instance, is the possibility of autocatalytic effects which presents probably the most difficult assumption made in the model. Since the calculation of autocatalytic effects within dynamic systems necessarily leads to another level of systems complexity, which at this stage of model development would result in basic parameters outlined here conflicting with derived parameters (containing feedback loops), generally making DMM more difficult to grasp autocatalytic effects are ignored. It is hoped that research on TLA, especially on automaticity, will contribute to the understanding of the autocatalytic effects in multilingual systems.

Therefore the following factors which have intentionally been excluded from the model, such as

- the idea of a qualitative change within a system,
- the idea of synergetic effects of language systems and
- the idea of the development of dispositional and emergent properties unique to multilingual systems

will be paid particular attention to now.

The dependency models developed include the possibility of feedback loops, which can be seen as representing possible autocatalytic effects. As anyone who has dealt with biological growth models will know, a specific parameter can be 'set off' by outward influences and go into a 'growth mode' without any further external influences. This growth mode might be retarded by lack of external input but will in a certain phase have to be described as autocatalytic, that is relying on no external cause. This kind of growth is at least at present not predictable in DMM, although DMM does operate with critical stages or thresholds. The introduction of a critical stage only names the phenomenon, but does not explain it. Qualitative changes are well-known in systems theory, but a theory of qualitative change in psycholinguistics must appear fairly novel.

Furthermore DMM includes separate components for the language systems involved and treats them as inherently autonomous. Firstly, the development of a language system outlined in DMM depends on the development of its language components. It is clear that the development of the subcomponent (1) will depend on the state of the subcomponent (2) and therefore language growth, for example, will take place in more uneven spurts than predicted by DMM. We are also aware of the phenomenon of overgeneralisation leading to transitional setback in language development due to the restructuring of the language systems (see

McLaughlin, 1987: 123–124). This phenomenon observed in the course of LS restructuring is also known as U-shaped behaviour (see Sharwood Smith & Kellerman, 1989: 221). It is important to realise that the growth predicted by DMM is an average of sudden growth set off by intermittent setbacks and not a realistic prediction of a smooth language development. Secondly, according to DMM, LS_1, LS_2, LS_3 and LS_n are dealt with as if they were identical systems. Whereas UG theory assumes that all natural languages share certain properties which, if learned by the speaker, will lead to synergetic effects in language acquisition DMM makes no predictions in this respect.

One claim made by DMM is that the need to deal with more than one language in a multilingual system will lead to the development of certain skills or dispositional properties in the language speaker which will facilitate the acquisition and maintenance of further language systems. This can in a sense be described as a meta-autocatalytic effect. Emergent properties which develop as a result of such an effect 'cannot be calculated by adding up the properties of the components but result from the interaction of the components' (Strohner, 1995: 29).

Note that we distinguish emergent properties from systems properties, in so far as systems or gestalt properties are those of the whole system not attributable to any part of the system (the ability of a plane to fly is not attributable to any part of the given system). Emergent properties are on the other hand only to be found in open systems, in so far as these properties are such that they will, or are likely to develop in a system but are not systems properties per se (see also Stadler & Kruse, 1992).

Whilst it may be clear that systems properties can only be discussed in a systems-theoretic interpretation of multilingualism, it is important to note that the individual factors explained are only significant in so far as they contribute to a new overall picture, like pieces of a puzzle. And it is to this overall picture that we shall now turn.

Chapter 8
Holism Defended: A Systems Interpretation

As has already been stated, it became obvious at an early stage in research on multilingualism that the measurement of bilingualism in terms of how far the multilingual person manages to achieve the level of monolingual competence is highly misleading (see Keatley, 1992: 52). As argued by Grosjean and the authors of this book, the bilingual speaker is not a double monolingual but the multilingual speaker develops cognitive and linguistic skills that differ substantially from those required by and expected of the monolingual speaker. Multilingualism appears to effect substantial cognitive and linguistic changes in the speaker which force upon the linguist investigating the phenomenon of multilingualism a view of language competences which must be described as holistic.

The most obvious solution to the problems raised by research on multilingualism, which necessarily affects the way we view language skills as a whole, is a modular one. The concept of a unitary language capacity termed language competence was attributed certain properties such as innateness and cognitive independence, has been replaced by a more modular conception of language competence. This is, however, frequently used as a means to sidestep the basic problem of holism. That is, even if one is prepared to give up the concept of a homogeneous and unitary competence (as discussed in Chapter 5), a modular approach can be used as a stop-gap to deal with the evident inadequacies of a unitary approach to language competence. A modular interpretation of unitary competence consists of a modularisation of the necessary skills, as originally envisaged by Chomsky.

In this chapter we deal with the holistic approach taken in DMM first by discussing why we reject a modular view of language competence according to a systems-theoretic view and second by comparing the approach taken in DMM to the wholistic concepts of bilingualism as proposed by Grosjean and Cook.

Double Monolingualism and Modularity

It is generally agreed that language-related mental functions are modular by nature. Ever since the publication of Fodor's *The Modularity of Mind* (1983), modularity has been one of the assumptions of psycholinguistic research on language acquisition.

Before we discuss modularity as such, it will be important to distinguish between two types of modularity claims which we shall term the weak and the strong modularity claim. Weak modularity as postulated by Chomsky is described by Salkie (1990: 87): '[T]he mind is modular by nature. Each 'module' is a discrete part of the mind, with its own structure and organising principles, in actual behaviour, the different modules of the mind interact, but it is none the less possible – indeed, necessary to distinguish them.'

What Salkie, however, does not mention is that one could interpret Chomsky as making a strong modularity claim. That is to say, the early definition of the language acquisition device and subsequent research by UG-oriented psycholinguists would suggest not merely the modularity of mind but the autonomy of the specific modules. Thus a frequent claim is that of the cognitive independence of language functions, to which there is a tradition of research into mental retardation and language competence (see Smith & Tsimpli, 1995; Skehan, 1998: 207–235).

The problem of claiming UG to be universal on the one hand and coming to terms with the fact that language impairment is a common phenomenon (more evident in non-native speakers) on the other hand has provoked a number of ingenious explanations stemming from proponents of UG interpretations of the language acquisition process. As already mentioned before in Chapters 4 and 6, one of the most recent analyses is that of Christopher, an interesting case of a language savant with a partial knowledge of sixteen foreign languages, as provided by Smith and Tsimpli (1995). Bates (1997: 171–172) critically comments on their approach:

> In their application of UG to C's [Christopher's] data and that of normal controls S&T [Smith & Tsimpli] have added some additional mechanisms that protect the theory from disconfirmation, in a complex variant of the old competence/performance distinction. They start out with a learning model (a hybrid of Fodor, 1983 and Anderson, 1992) in which the language module is encapsulated from central processing, but they end up with a model in which intimate interactions between UG and the general learning mechanism are permitted throughout the language learning process. The GLM [general learning mechanism] now serves as a rather fickle *deus ex machina*, rescuing the L2 learner

from old parameter settings in some situations [...] but permitting L1 structures to sneak into L2 on others [...].

Weak modularity is easily integrated into DMM, as in DMM we also assume that there must exist various distinct components contributing to observed language performance. In view of the fact that we know that these modular functions do not have invariable neurophysiological correlates weak modularity appears to be an organisational necessity and does not represent a problem for DMM.

Strong modularity on the other hand would claim the independence of these modular functions. This was also originally suggested by Chomsky, who claimed that the language acquisition device was responsible for language acquisition and that this was not dependent on other cognitive functions. This claim was supported by research on cognitively deficient individuals whose language competence seemed unimpaired (see the case of Marta cited by Aitchison, 1989: 146).

The concept of strong modularity is neatly outlined by Gardner (1983: 56):

> Though, again, there are differences, most exponents of a so-called modular view are not friendly to the notion of a central information processing mechanism that decides which computer to invoke [...]. Nor is there much sympathy for the notion of a general working memory or storage space which can be equally well used (or borrowed) by the different special-purpose computational mechanisms. Instead the thrust of this biologically oriented position is that each intellectual mechanism works pretty much under its own steam, using its own peculiar perceptual and mnemonic capacities with little reason (or need) to borrow space from another module.

Strong modularity must be taken to have at least two consequences:

(1) If the modular functions are independent, the transition from explicit language knowledge to implicit language knowledge (as suggested by the ACT model) is not feasible – a line taken by Krashen (1981).
(2) If the modular functions are independent, cognitive functions should not be able to restrict the working of purely linguistic functions as suggested by DMM. It is significant that Smith and Tsimpli's investigations (1995) lead to the assumption that Fodorian modular conception of mind is inappropriate as a 'central system' appears to be a necessary component in multilingual language management and Christopher appears to be deficient in this aspect in particular as already pointed out before. Christopher's multilingualism must

therefore be considered severely impaired. This assumption is confirmed by the results obtained by Smith and Tsimpli.

As already pointed out in Chapter 4, we would like to note that the investigations conducted by Smith and Tsimpli are one of the most interesting pieces of research recently conducted on the question of modular ability and some of the results are sufficiently unusual to require further investigation. In their afterword the authors themselves suggest that '[a]s well as its implications for linguistic and pragmatic theory, Christopher's case has potential, but somewhat more problematic, implications for the theory of second language acquisition.' Despite the fact that the researchers argue for a (modified) modular structure of mind, they admit that Christopher's performance is impaired by the phenomenon of 'cognitive overload' (Smith & Tsimpli, 1995: 182), which in view of the competence orientation of UG based research is an understandable attribution. The results presented so far are compatible with the cognitive functions specified in DMM.

Before we turn to the wholistic concepts of multilinguals introduced by Grosjean and Cook, we would like to touch upon the neurophysiological arguments for modularity. Within the UG tradition it has been customary to assume that specific functions derived from the UG model would have to be localised in a certain device. In the course of time the language acquisition device was complemented by a plethora of other devices taking care of their respective functions. In the best of Cartesian traditions these devices were then anatomically localised, or at least attempts were made to determine a corresponding neurophysiological localisation by experimental or biographical evidence. As plausible as this may seem, these attempts are based arguments similarly fallacious to those which led Descartes to place the soul in the pituitary gland.

Although we do not doubt the existence of an onto- and phylogenetically determined differentiation of language functions, we still consider the assumption that every function has to have a neurophysiological equivalent incorrect.

- There is ample evidence to show that the brain is essentially a parallel processing system, which means that even at a later stage of neurophysiological differentiation (with resultant loss of processing flexibility) the same function can be obtained by different neurophysiological processes. This claim is supported by the redundancy of information storage in memory. For every function there is therefore more than one possible neurophysiological interpretation (see Churchland & Sejnowski, 1990: 224–227).

- According to our present state of knowledge experimental or biographical evidence for the localisation of particular functions is at best inconclusive, if not misleading. In the analysis of empirical evidence we must clearly assume that the 'language function' is a complex one, and does not depend exclusively on the presence or absence of one particular property.
- As illustrated by the previous examples, complex properties or functions must be seen as systems properties and not properties attributable to a certain systems component. From a holistic or systems-theoretic point of view we can call this the componential fallacy of functions. As argued by Strohner (1995: 48) there is a basic fault in modularity theory which sees the human brain in analogy to a computer constructed according to von Neumann principles: 'Wie wir heute aufgrund neuer Untersuchungsmethoden immer genauer wissen, hat die Natur dem Gehirn einen völlig anderen Bauplan zugrunde gelegt als die Informatiker einem Standardcomputer.' (As we know today, on the basis of new methods of investigation nature has based the brain on a completely different design than computer engineers have the common computer. Translation by the authors)

A careful investigation of the strong modularity claim shows that this working hypothesis does not comply with the principle of neuronal adequacy. The most substantial argument against modularity must be that the modularity argument explicitly or implicitly presuppposes a view of mind as mechanism. Insights gained by neuroscience have, however, increasingly exposed the severe limitations of a mechanical view of mind (e.g. see, Springer & Deutsch, 1981).

Wholism: The Bilingual View and Multicompetence

As Grosjean has argued and has been pointed out in previous chapters the double monolingualism view is inadequate in the attempt to develop a realistic model of multilingualism. Even if this view is not necessarily explicitly adopted by UG-oriented language acquisition research, the discussion centring on the accessibility of UG is certainly governed by an implicit double monolingualism conception of multilingualism. This interpretation coincides with the observations of Ellis (1994) and Cook (1993a: 244). Illitch and Sanders (1988: 52) also point out that:

> [o]ne obstacle most modern readers face when they want to study the history of 'language' is their belief in monolingual man. From Saussure to Chomsky 'homo monolinguis' is posited as 'the man who uses

languages' […]. [L]inguistic theories and descriptions have ultimately to account for the multilingual nature of most people's knowledge of language; principles and parameters cannot be expressed in such a form that it is impossible for one mind to hold more than one grammar at a time.

If the double monolingualism view is entirely inappropriate, the question obviously arises, which alternative models can be used in multilingualism research. The two models most commonly referred to are those suggested by Grosjean and Cook.

Grosjean (e.g. 1985, 1998) is the first to be clear on the rejection of the double monolingualism view and he also dismisses switch models. He appears to favour a unitary conception of multilingualism (i.e. the non-reducibility to a double monolingual view of multilingual competence) based on an interactive activation view of multilingual language management. According to Grosjean it must be assumed that the bilingual speaker will activate the language network required by the specific communicative situation: the speaker will thus opt for a base language required by the communicative situation, which will thus represent her/his primary activated language system:

> When the bilingual is in a bilingual speech mode, both language networks are activated but the base language network is more strongly activated; the activation of a unit (e.g. phoneme, syllable, word etc.) in one network and of its 'counterpart' in the other (if it exists) depends on their degree of similarity; the activation of units specific to one language increases the overall activation of that language network and thus speeds up the recognition of words in that language; finally, the activation of words that are similar in the two lexicons will normally slow down the recognition of guest language words. (Grosjean, 1992: 761)

Cook's model of multicompetence is by his own definition also a wholistic one. On a simple level this must be taken to mean that multicompetence is unitary. On a more complex level this must be taken to mean that the language systems as such do not represent separate systems but one system. That is, the bilingual speaker of French and English does not have command of the language systems French and English but of a unitary language system we would have to call French-English, as Franglais is already taken to mean something else.

DMM on the other hand takes a h(!)olistic approach to multilingualism. Whilst DMM assumes that multilinguals cannot be measured by monolingual standards and that multilingualism cannot simply be explained by extended monolingual acquisition models, DMM also claims that the language systems involved can be interpreted as separate systems. They are also perceived as such by the multilingual speaker, although s/he might admit to difficulties keeping them apart. This means that on the one hand the dynamic model is separatist or modular in interpreting the involved language systems and factors as separate modules, but assumes on the other hand that the subsystems outlined interact with each other and influence each other within the complex and dynamic system we call multilingualism. This approach complies with the suggestions made by Singleton (1996: 250–251) although he focuses mainly on the bilingual lexicon: 'The message which emerges from the foregoing is that neither a strict separatist nor a strict integrationist model of bilingual lexical organisation is supported by the available evidence.' How the holistic approach taken by DMM differs from the unitary approaches outlined above should become more evident in the following.

Holism and Systems Theory

Like Cook, DMM takes a necessarily wholistic view of multilingualism. This should, however, not merely be understood as meaning that the multilingual system is taken as a speaker system in its own right, but that this system should be modelled according to holistic principles. A simple presentation of these holistic principles is to be found in Phillips (1992: 724), in which he outlines four characteristic working hypotheses of a holistic approach:

- the whole is more than the sum of its parts;
- the whole determines the nature of its parts;
- the parts cannot be understood, if considered in isolation from the whole;
- the parts are dynamically interrelated or dependent.

These characteristics are basic presuppositions of a holistic view of any object of investigation. Such a holistic view is a necessary presupposition of a dynamic view. Holism as such is, however, only a minor aspect of the development of DMM. Only when we have managed to specify the relations between the various components of the system and specified how these relations affect other parts of the system, can we begin to model the system in time, that is let the system develop. A holistic approach will provide us with a system and a dynamic system will have to result from the specification of the interrelations between the parts of the system. Such an approach can best be specified as analytical holism.

From the ongoing discussion it has become clear that we use both the term 'wholistic' and 'holistic' but with a subtle difference. Whereas 'wholistic' expresses the preparedness to view the phenomena observed as a whole and not merely its parts, 'holistic' refers to a specific theoretical position, which, for example, assumes that systems as a whole will have properties their parts cannot be shown to contain.

The introduction of a systems metaphor into multilingualism research might be seen as an innovative and even useful approach. But to describe something as a system as such is probably too undifferentiated a judgement to be of much use. As pointed out by von Bertalanffy (1968) and Waddington (1977), the systems view can be applied to most objects of investigation. The very much more important question is of which type the multilingual system is. According to DMM the (psycholinguistic) multilingual system is dynamic and adaptive. The multilingual system is accordingly characterised by continuous change and non-linear growth. As an adaptive system it possesses the property of elasticity (ability to adapt to temporary changes in the systems environment) and plasticity (ability to develop new systems properties in response to altered environmental conditions) (see Strohner, 1995: 32–33). All these properties are illustrated by the graphs provided.

DMM aims to provide the model of such a system for multilingualism. To avoid unnecessary complication, the (systems) variables had to be introduced gradually. Only after satisfactory acquaintance had been made with the concept of transitional competence, only after the specification of the independent and dependent variables and their relations can we procede to specify conditions of individual variations. This necessarily means that DMM started out being an excessively abstract model (obviously open to a large number of objections based on empirical evidence), which, however, constituted a necessary precondition for the derivation of interrelations between variables and individual language development. Obviously, in many instances the relationships postulated by DMM could only be sketched – to give a general impression of the variable development to be expected – and would require confirmation through detailed empirical research. In many instances the measures for particular skills, dispositions and effects have yet to be developed.

One of the impressions obtained is that research on multilingualism is still in its infancy and most of the work still lies ahead. We hope that DMM will provide an innovative theoretical framework in which it is possible to ask meaningful questions concerning multilingual development and obtain more satisfactory answers to the plethora of questions surrounding multilingualism as a psycholinguistic phenomenon with sociolinguistic consequences. One thing that should have become clear is that the UG model is in many respects a hindrance rather than a help in developing a realistic view of multilingualism and that a realistic view requires that we throw overboard many a past insight provided by language acquisition theory, as it has evidently outlived its usefulness.

Chapter 9
Limitations, Conclusions and Outlook

Although we are convinced that DMM provides a useful model of multilingualism we would like to address some theoretical problems of heuristic models such as DMM in this final chapter of the book. During the last years several studies applying a systems-theoretic approach to language learning have been published and this supportive evidence for the plausibility and popularity of our approach will be described in a short overview. We will also pose a few of the many questions raised in the discussion of the model which had to be left open for the time being and thus present suggestions for future research on multilingualism. Future perspectives of multilingualism will also be presented in our ideas about the application of some concepts developed in our realistic model to multilingual education at the end of the book.

Theoretical Limitations of DMM

It would be only too easy to misunderstand the claims made by DMM. On the basis of recent insights gained in the theory of science to the effect that all our theories should be seen as models and all our models are in a way metaphors, the empirical claims made by DMM are decidedly weak. The usefulness of DMM must lie not in the fact that it provides better explanations of empirical evidence but rather that it provides a convincing and useful model or metaphor which should enable us to think about a multitude of seemingly contradictory and confusing phenomena related to multilingualism in a more coherent and cogent way. This understanding of DMM explains why it exhibits the disadvantages of most heuristic models.

The modelling of DMM presupposes a certain amount of insight into the nature of scientific thinking. An investigation of recent developments in the methodology of sciences shows that the empiricist-inductivist approach to scientific research represents a severe distortion of effective scientific methodology. Post-Popperian methodology will definitely assume that scientific research presupposes the prior existence of working hypotheses, which in effect determine the relevance and interpretation of empirical evidence available. Hypothesis formation therefore by necessity precedes rather than follows empirical research. More recent non-statement theoretical methodology has established convincing arguments for the function of theory being the provision of formalisable models of reality and not statements about reality as such. The empirical content of a theory therefore results from a specific interpretation of the theory in the form of the application to a specific segment of the world to which the theory is considered applicable (see Fetzer, 1993 and Boyd & Gasper & Trout, 1991). The attempts to explicate the shortcomings of current linguistic methodology can be found in Herdina (1990, 1992a, 1996, 1997).

Another aspect to be considered is the theoretical framework in which DMM is conceived. DMM works on the assumption that human beings are complex biological systems and therefore are best studied and explained on the basis of systems theory, complexity theory and chaos theory rather than fundamentally mentalist and methodological models. For an early criticism of the misapprehension of man in such models the reader is referred to Herdina (1988). The conventional view of how the mind, society and the world works appears to have run into a crisis or worked itself into a blind alley and is rapidly being replaced at least in some spheres (including economics) by a new mode of thinking with a completely new set of concepts and understandings. Fortunately, we are presently able to observe a paradigm shift taking place, favouring a systemic and holistic view of both humans and their societies (see Shore, 1995: 104).

This position is most clearly set out in Eisenhardt, Kurth and Stiehl (1995: 2), whose book begins with an explicit statement, which we have translated as follows:

> Modern science and technology are based on two illusions: the assumption that there is some global truth and the assumption that reality is static and consists of independent elements, which are compartmentalised and do not interact. The theory of complex systems (synergetics, chaos theory), the theory of fractals and the theory of self-organising automata: they all require, even force upon us a new view of science and the world. The only thing that is, is a finite, localised and interactive process.

As a new model DMM does not merely provide minor corrections to established theories but suggests a whole new view of language-related processes and much of its theoretical attractiveness must lie in the plausibility of the model as a whole. Why it makes more sense to take a systems-theoretic view of language processes is something that cannot be logically proven on the basis of a single argument. Methodologically DMM contains insights into the workings of science which can only be explicated to a minimal extent here and we therefore hope that our goal to provide a satisfactory outline of the systems-theoretic presuppositions of DMM has been fulfilled. Fortunately, since the first mention of DMM in 1994 in a conference paper on the paradox of transfer (Jessner & Herdina, 1994) systems-theoretic approaches have become more popular in language-related research, too, and recent publications referring to dynamic systems within a linguistic context suggest that such a model is rapidly gaining popularity.

Concepts Related to DMM

In the following we would like to give an overview of those publications we consider related to our systems-theoretic approach. They focus on language acquisition theory, on cognitive aspects of language learning, neurological aspects of language learning and on bilingual lexical networks. We will also have a look at connectionism and new concepts such as emergentism. Applications of a systems-theoretic approach to language education will finally be addressed.

Tucker & Hirsh-Pasek (1993) consider language acquisition a problem of the emergence of new forms in a complex system. They identify two sorts of theories, that is inside-out theories and outside-in theories:

> The former are characterised by strong, usually innate, language-specific algorithms or heurististics which propel the child forward in the acquisition task (e.g. Chomsky, 1981; Pinker, 1984; Schlesinger, 1987). The latter class of theories is characterised by the belief in a highly structured environment, and the presence of domain-general learning abilities (e.g. sensitivity to salience, inductive capacity, category formation) to account for syntactic competence in the child (e.g. Bates et al., 1979; Bates & Mac Whinney, 1987; Snow, 1989).

They point out that this theory-dichotomy represents polar extremes on a continuum, and most theories fall somewhere between them, and that they characterise the contributions of environmental and biological or innate factors as standing in an additive, linear relationship to one another. Furthermore most theories concentrate on one of the three linguistic subsystems, mainly syntax, in explaining the development of adult competence, while failing to recognise the influence of other systems. In their approach

> there is no attempt to appeal to the existence of information either in the environment or in the individual, as innate structure, to account for development. Structure or form (information) is *constructed* in

development, and arises through the successive organisational adaptation of systems components to a specific context. Second, a central tenet in systems theory is the inherent organization and interdependence of systems components, and the progression from lower to higher, more complex levels of organiaztion in development. The theory predicts that the linguistic subsystems interact in complex ways, and that the relative contribution of each subsystem will change with development. (Tucker & Hirsh-Pasek, 1993: 362)

With the help of a dynamic systems approach they try to provide a reinterpretation of language acquisition theory concerning the acquisition of first generation creole languages from contact pidgins which they see as self-organising, children's apparently rule-driven errors which they identify as an unstable dynamic state and the development of idiosyncratic linguistic forms which also show self-organising character. More recent work of these authors is included in MacWhinney (1999) which will be addressed later.

Lightfoot (1999) attempts to combine child language, the history of English and evolutionary biology in his approach to language development. As already pointed out in Chapter 4, he presents arguments for an innate language faculty – thus following Chomsky – but does not consider it necessarily a unique faculty but one of a number of mental abilities.

In his interdisciplinary linguistic attractor model Cooper (1999) describes language processing as fractal sets and investigates the dynamics of the fractals in the individual and the social level of the speech community. He introduces techniques to isolate and measure attractors with regard to stability and applies his approach to phonetic change and metrics.

Van Geert (1993) introduces the ecological metaphor to cognitive growth and speaks about a cognitive grower and its environment in his dynamic systems approach to cognitive development. He defines cognitive growth

> as an autocatalytic quantitative increase in a growth variable, following the emergence of a specific structural possibility in the cognitive system. Examples include the growth of vocabulary, the growth of subject–verb inversion in interrogative sentences, the growth of the correct use of a strategy in solving fractions problems, and the like. (van Geert, 1993: 274–275)

In his book (1994) he focuses on dynamic systems of (monolingual) language development in children.

Another strand of research draws on the principle of self-organisation in neural networks in language learning (e.g. Karpf, 1990; Peltzer-Karpf, 1996; Zangl, 1998). According to this theoretical framework, which sees modularity as a developmental principle neurological and functional aspects are closely differentiated. The studies mainly concentrate on the development of morpho-syntactic and lexico-semantic aspects in (foreign) language learning.

In his studies on lexical networks Meara (e.g. 1997, 1999) presents a new approach to modelling vocabulary acquisition by drawing on self-organisation. His work can be seen complementary to ours; in Meara (1999), for instance, he refers to the unexpected activation of prior language knowledge in a second or third language context. In his explanations he draws on Random Boolean Networks to introduce systems-theoretic principles and he presents bi- and trilingual networks in his simulations. He provides stimulating insights into the way real mental lexicons in multilinguals work.

Connectionism (e.g. Rumelhart & McClelland, 1986) which has established itself as a new approach to language learning can be contrasted with a cognitivist view. According to connectionists knowledge as a command of a language is seen as an emergent property of a system of atomic elements between decision nodes. The cognitivist view which is based on the assumption that the mental system required for language has to be a symbolic system with rules is thus replaced by the idea of knowledge or rule-governed behaviour as the result of learning. Researchers working with the new computer metaphor or brain metaphor, which is variously known as connectionism, parallel distributing processing and/or neural networks, are interested in the study of change, that is changes in behaviour, changes in the neural structures that underlie behaviour, and changes in the relationship between mind and brain across the course of development (see Bates & Elman, 1992: 2). They assume that our cognitive system works like a network of interconnected units. Parallel distributing then refers to the assumption that different parts of information are processed independently of one another on different levels.

Emergentism (Elman *et al.*, 1996; MacWhinney, 1999) has arisen as a new approach to language acquisition which stands in close relationship to the connectionist model. Unlike nativism emergentism sees language as emerging from interactions between biological and environmental processes. Emergentism also operates with principles from systems theory viewing language as 'a dynamic, complex, non-linear system where the timing of events can have a dramatic influence on the developmental course and outcomes.' (Ellis, 1998: 642). Ellis (1998: 645) states that emergentism needs the computational tools provided by connectionism to explore the conditions under which emergent properties arise.

The application of a complexity or chaos-theoretical approach can be found in Bleyhl (1997) and Larsen-Freeman (1997). Bleyhl provides a critical article on the state of art in foreign language teaching. He discusses the dynamics and non-linear development in language learning suggesting a new approach to language learning in the classroom building on chaos theory. Larsen-Freeman relates chaos theory in a very general way to research on SLA by pointing to the advantages such an approach could have for research in applied linguistics. It is to be noted that we do not merely see dynamic modelling as a useful analogy as does Larsen-Freeman (1997: 141), who claims that 'the analogy might only be metaphoric,' whilst we do agree with her statement that 'the traditional approach to science which attempts to understand the behavior of the whole by examining its parts piecemeal is inadequate for studying complex systems.'

Questions Raised by DMM

Although the authors believe that DMM provides a suitable explanatory framework to integrate the majority of the phenomena observed in multilingualism research, the dynamic approach raises a number of questions that have been left unanswered, some of which are listed below.

- If language learning and forgetting processes are not linear but appear to obey abstract mathematical principles of growth and loss, what are the actual physical processes going in the multilingual brain that make such processes possible?
- What causes autocatalytic learning processes to occur at an object level? Can we conceive of them as recursive cognitive feedback loops which lead to dynamically incremental growth of available language knowledge?
- What kind of metaknowledge does second language learning lead to and more significantly, does this knowledge conform to the same principles of cognitive growth as the underlying cognitive growth processes? Does this mean that TLA obeys essentially different principles than those governing SLA?
- Is there something like a command of a language in some sort of stable way (commonly referred to as competence) or is this a hidden variable? Our answer to this question will probably – at least to some extent – depend on whether we wish to view language competence as related to something like explicit knowledge or whether we assume

competence to describe a type of knowledge that is in essence implicit and procedural.
- How can language maintenance be measured?

In sum, we would thus like to stress the importance of further research on the cognitive aspects involved in language processing and language learning as well as language loss or attrition. Another focus of interest will be TLA (and fourth language acquisition) and their relation to SLA.

One of the most important aspects of language acquisition as identified by DMM is language maintenance (effort) which will hopefully receive more attention by researchers in the future. Psycholinguistic research has already begun to focus on language retention (see, e.g. Beaton & Gruneberg & Ellis, 1995) but more work is needed in order to gain valuable insights into the processes involved in the maintenance of a language system in a multilingual individual.

Not only does DMM appear to fit into a general trend to take multilingualism far more seriously as a key issue of (applied) research, there are also a number of issues DMM throws a new – and we hope – clarifying light on.

Applications of DMM

One of the most challenging problems of applied linguistics is multilingual education. Multilingualism on the individual level as the goal of multilingual education is a complex ability that cannot be provided by mereley adding more language subjects to the curriculum. A realistic model of language acquisition such as the DMM is aimed at adding to a better understanding of the on-going processes in language acquisition and at helping to overcome the belief that bilingualism is nothing more than a handicap. Thus DMM could serve as a missing link between language education and linguistics in the development of new language teaching concepts, that is, to provide a psycholinguistic basis for applied research. The need for a re-evaluation is also clearly stated by Larsen-Freeman (1997: 152): 'Seeing language as a complex non-linear system may cause us to re-evaluate our assumptions about the basic mechanisms […] operating in SLA.'

Our dynamic model stresses features of modern language education that have been postulated by educationists but so far have not been supported by language research. In the following we would like to point out some of the salient features of plurilingual (and multicultural) education as, for instance, discussed by Coste (1997) and link them to some of the main ideas taken in DMM to demonstrate the potential such a model has to offer for educational needs.

Coste (1997) draws a complex picture of a plurilingual and pluricultural competence serving as the goal of future multilingual education at school. He points out that this picture is based on a multilingual view compared with the representation of the ideal communicator. He characterises the unevenness of plurilingual and pluricultural competence by pointing out that the multilingual learner usually shows greater proficiency in one language as compared with the others and a different profile of competences in one language as compared with others. For example, excellent speaking competence in two languages, but good writing competence in only one of them, with a third language being only mastered as regards written comprehension, without any real oral ability. According to Coste (1997: 91) the dynamics on the individual level are linked to 'occupational, geographical and family movements and of changing personal interests.' Furthermore he considers the concept of partial competence in a particular language meaningful in such a way that it forms part of a multiple competence which it enriches. He also stresses the importance of metalinguistic knowledge and skills and their facilitative nature for further language learning.

This selection of ideas on the dynamic nature of multilingual proficiency, approximative systems and the M-factor was chosen to illustrate that the psycholinguistic basis as provided in DMM could back up the goals of multilingual education. Such a holistic model supporting both the view that language proficiency (competence) is not a steady state and that having to deal with more than one language at a time does not mean that a person's cognitive resources are divided and therefore reduced could serve as the linguistic model needed to be able to understand the complexity of the educational problems of multilingualism.

This complexity is related to the qualitative changes in the speaker's psycholinguistic system. The development of multilingual proficiency leads to an enrichment of the individual language system but, as the whole system adapts to new environmental and psychological communicative requirements as perceived by the speaker, also changes its nature. As a result of these changes the multilingual repertoire develops new skills which fall into the categories of language learning skills, language management skills and language maintenance skills.

It is suggested here that these three skills, identified by DMM as part of the M-factor, should be focused on in future language education – preferably on all levels. Due to the contact with more language systems enhanced metalinguistic and metacognitive awareness develop in multilinguals and it is suggested that the instruction of such metaknowledge should be an integrative part of language education. For instance, metacognitive knowledge concerning language learning skills might also be an issue addressed in the classroom, that is, apart from the teacher who is supposed to provide support with language learning strategies the students could be encouraged to report on the strategies already employed in prior language learning. The reactivation of prior language knowledge in the classroom, that is to build on language systems which are already existent in the students' multilingual repertoire, is suggested to be of a facilitative nature in language learning. And at the same time metalinguistic and metacognitive awareness will be fostered in the students as argued by Jessner (1999; 2001). In particular, the focus on the similarities between two language systems (or cognates) as known from studying the behaviour of trilingual subjects seems to be helpful in the language learning process and it is therefore also pointed at the linguistic and educational advantages gained from bridges built by the teacher between otherwise isolated language subjects.

Additionally in the future language maintenance has to be reserved a far more bigger part in the classroom. According to DMM it seems advisable to concentrate more on language maintenance work in order to support progress in language acquisition. Apart from instructing teachers on the important role of language maintenance this involves focusing on testing methods that are to be developed and teaching materials concentrating on language maintenance skills. This also refers to material dedicated to the increase of metalinguistic awareness or language awareness in the multilingual classroom. Due to the language awareness movement within the European Community some of the latter nevertheless can be reported to be already on its way (see, e.g. Feichtinger *et al.*, 2000; Nagel, 2000).

There is also a further issue that should be touched upon. Research on multilingualism is not merely a crucial issue in the development of linguistic research. Multilingualism as such is obviously also a significant political issue. A useful effect of the adoption of a dynamic model of multilingualism might well be the realisation of the need to revise politically comfortable, but scientifically doubtful prejudices and to replace these by a more differentiated view of language acquisition and language use in a multicultural and multilingual (global) society.

References

Adjemian, C. (1976) On the nature of interlanguage systems. *Language Learning* 26, 297–320.

Aitchison, J. (1981) *Language Change: Progress or Decay?* Bungay: Fontana.

Aitchison, J. (1989) *The Articulate Mammal. An Introduction to Psycholinguistics.* 2nd edn. London: Unwin.

Albert, M. and Obler, L. (1978) *The Bilingual Brain: Neuropsychological and Neurolinguistic Aspects of Bilingualism.* New York: Academic Press.

Altenberg, E. (1991) Assessing first language vulnerability to attrition. In H. Seliger and R. Vago (eds) *First Language Attrition* (pp. 189–207). Cambridge: Cambridge University Press.

Anderson, J. (1983) *The Architecture of Cognition.* Cambridge, MA: Harvard University Press.

Anderson, J. (1995) *Cognitive Psychology and its Implications.* San Francisco: Freeman.

Appel, R. and Muysken, P. (1987) *Language Contact and Bilingualism.* London: Edward Arnold.

Arnberg, L. (1987) *Raising Children Bilingually: The Preschool Years.* Clevedon: Multilingual Matters.

Aslin, R. (1993) Commentary: The strange attractiveness of dynamic systems to development. In L. Smith and E. Thelen (eds) *A Dynamic Systems Approach to Development: Applications* (pp. 385–399). Cambridge, MA: MIT Press.

Auer, P. (1995) The pragmatics of code-switching. In L. Milroy and P. Muysken (eds) *One Speaker, Two Languages: Cross-disciplinary Perspectives on Code-switching* (pp. 1–24). Cambridge: Cambridge University Press.

Bachman, L. (1990) *Fundamental Considerations in Language Testing.* Oxford: Oxford University Press.

Bachman, L. (1991) What does language testing have to offer? *TESOL Quarterly* 25 (4), 671–704.

Bachman, L. and Palmer, A. (1982) The construct validation of some components of communicative proficiency. *TESOL Quarterly* 16 (4), 449–465.

Baker, C. (1996) *Foundations of Bilingualism and Bilingual Education.* Clevedon: Multilingual Matters.

Baker, C. and Prys Jones, S. (1998) *Encyclopedia of Bilingualism and Bilingual Education.* Clevedon: Multilingual Matters.

Bailey, C. (1973) *Variation and Linguistic Theory.* Washington, DC: Center for Applied Linguistics.

Barron-Hauwaert, S. (2000) Issues surrounding trilingual families: Children with simultaneous exposure to three languages. In J. Cenoz, B. Hufeisen and U. Jessner (eds) Trilingualism – Tertiary Languages – German in a multilingual world. Special Issue of *Journal of Intercultural Learning* 5 (1). On web at http://www.ualberta.ca/~german/ejournal/ejournal.htm

References

Basieux, P. (1995) *Die Welt als Roulette. Denken in Erwartungen*. Reinbek bei Hamburg: Rowohlt.

Bates, E. (1997) On language savants and the structure of the mind. Review of: *The Mind of a Savant: Language Learning and Modularity*, by Neil Smith and Ianthi-Maria Tsimpli, 1995. *International Journal of Bilingualism* 1 (2), 163–186.

Bates, E. and Carnevale, G. (1992) Developmental psychology in the 1990s: Language development. Project in Cognitive Neurodevelopment. CRL Technical Report 9204. Center for Research in Language, University of California, San Diego, May 1992.

Bates, E. and Elman, J. (1992) Connectionism and the study of change. CRL Technical Report 9202. Center for Research in Language, University of California, San Diego, February 1992.

Bateson, G. (1972) *Steps to an Ecology of Mind: Collected Essays in Anthropology, Psychiatry, Evolution and Epistemology*. Frogmore: Paladin.

Bateson, G. (1979) *Mind and Nature: A Necessary Unity*. New York: Dutton.

Beaton, A., Gruneberg, M. and Ellis, N. (1995) Retention of foreign vocabulary learned using the keyword method: A ten-year follow-up. *Second Language Research* 11 (2), 112–120.

Ben-Zeev, S. (1977) Mechanisms by which childhood bilingualism affects understanding of language and cognitive structures. In P. Hornby (ed.) *Bilingualism. Psychological, Social, and Educational Implications* (pp. 29–55). New York: Academic Press.

Bhatia, T. and Ritchie, W. (1996) Bilingual language mixing, Universal Grammar, and Second Language Acquisition. In W. Ritchie and T. Bhatia (eds) *Handbook of Second Language Acquisition* (pp. 627–689). New York: Academic Press.

Bialystok, E. (1990) *Communication Strategies*. London: Blackwell.

Bialystok, E. (1991a) Introduction. In E. Bialystok (ed.) *Language Processing in Bilingual Children* (pp. 1–9). Cambridge: Cambridge University Press.

Bialystok, E. (1991b) Metalinguistic dimensions of bilingual language proficiency. In E. Bialystok (ed.) *Language Processing in Bilingual Children* (pp. 113–140). Cambridge: Cambridge University Press.

Bialystok, E. (1992) Selective attention in bilingual processing. In R. Harris (ed.) *Cognitive Processing in Bilinguals* (pp. 501–514). Amsterdam: North Holland.

Bialystok, E. and Hakuta, K. (1994) *In Other Words: The Science and Psychology of Second-language Acquisition*. New York: Basic Books.

Bickerton, D. (1975) *Dynamics of a Creole System*. Cambridge: Cambridge University Press.

Birdsong, D. (1992) Ultimate attainment in second language acquisition. *Language* 68, 706–755.

Bley-Vroman, R. (1989) What is the logical problem of foreign language learning? In S. Gass and J. Schachter (eds) *Linguistic Perspectives on Second Language Acquisition* (pp. 41–68). Cambridge: Cambridge University Press.

Bleyhl, W. (1997) Fremdsprachenlernen als dynamischer und nichtlinearer Prozeß oder: weshalb die Bilanz des traditionellen Unterrichts und auch der Fremdsprachenforschung „nicht schmeichelhaft" sein kann. *Fremdsprachen Lehren und Lernen* 26, 219–238.

Bongaerts, T., Planken, B. and Schils, E. (1995) Can late starters attain a native accent in a foreign language? A test of the critical period hypothesis. In D. Singleton and

Z. Lengyel (eds) *The Age-factor in Second Language Acquisition* (pp. 30–50). Clevedon: Multilingual Matters.

Borer, H. and Wexler, K. (1987) The maturation of syntax. In T. Roeper and E. Williams (eds) *Parameter Setting* (pp. 123–172). Dordrecht: Reidel.

Boyd, R., Gasper, P. and Trout, J. (eds) (1991) *The Philosophy of Science*. Cambridge, MA: MIT Press.

Braitenberg, V. and Hosp, I. (eds) (1994) *Evolution. Entwicklung und Organisation in der Natur*. Reinbek bei Hamburg: Rowohlt.

Briggs, J. and Peat, F. (1989) *Turbulent Mirror: An Illustrated Guide to Chaos Theory and the Science of Wholeness*. New York: Harper & Row.

Brown, G. (1996) Language learning, competence and performance. In G. Brown, K. Malmkjaer and J. Williams (eds) *Performance and Competence in Second Language Acquisition* (pp. 185–203). Cambridge: Cambridge University Press.

Brown, G., Malmkjaer, K. and Williams, J. (eds) (1996) *Performance and Competence in Second Language Acquisition*. Cambridge: Cambridge University Press.

Canale, M. and Swain, M. (1980) Theoretical bases of communicative approaches to second language teaching and testing. *Applied Linguistics* 1, 1–47.

Carroll, J. (1981) Twenty-five years of research on foreign language aptitude. In K. Diller (ed.) *Individual Differences and Universals in Language Learning Aptitude* (pp. 83–118). Rowley, MA: Newbury House.

Carroll, J. and Sapon, S. (1959) *Modern Language Aptitude Test: Form A*. New York: The Psychological Corporation.

Casti, J. (1991) *Searching for Certainty: What Scientists Can Know About the Future*. London: Abacus.

Celce-Murcia, M., Dörnyei, Z. and Thurrell, S. (1995) Communicative competence: A pedagogically motivated framework with content specifications. *Issues in Applied Linguistics* 6, 5–35.

Cenoz, J. (2000) Research on multilingual acquisition. In J. Cenoz and U. Jessner (eds) (2000) *English in Europe: The Acquisition of a Third Language* (pp. 39–53). Clevedon: Multilingual Matters.

Cenoz, J. and Genesee, F. (1998) *Beyond Bilingualism: Multilingualism and Multilingual Education*. Clevedon: Multilingual Matters.

Cenoz, J., Hufeisen, B. and Jessner, U. (eds) (2001) *Cross-linguistic Influence in Third Language Acquisition: Psycholinguistic Perspectives*. Clevedon: Multilingual Matters.

Cenoz, J. and Jessner, U. (eds) (2000) *English in Europe: The Acquisition of a Third Language*. Clevedon: Multilingual Matters.

Cenoz, J. and Lindsay, D. (1994) Teaching English in primary school: A project to introduce a third language to eight-year-olds. *Language and Education* 8, 201–210.

Cenoz, J. and Valencia, J. (1994) Additive trilingualism: Evidence from the Basque Country. *Applied Psycholinguistics* 15, 197–209.

Chomsky, N. (1965) *Aspects of the Theory of Syntax*. Cambridge, MA: MIT Press.

Chomsky, N. (1977) *The Logical Structure of Linguistic Theory*. New York and London: Plenum Press.

Chomsky, N. (1980) *Rules and Representations*. New York: Columbia University Press.

Chomsky, N. (1986) *Knowledge of Language: Its Nature, Origin and Use*. New York: Praeger.

Chomsky, N. (1988) *Language and Problems of Knowledge: The Managua Lectures.* Cambridge, MA: MIT Press.
Churchland, P. and Sejnowski, T. (1990) Neural representations and neural computation. In W. Lycan (ed.) *Mind and Cognition* (pp. 224–252). Oxford: Blackwell.
Clahsen, H. and Muysken, P. (1986) The availability of universal grammar to adult and child learners: The study of the acquisition of German word order. *Second Language Research* 2, 93–119.
Clément, R., Smythe, P. and Gardner, R. (1980) Social and individual factors in second language acquisition. *Canadian Journal of Behavioural Science* 12 (4), 293–302.
Cline, T. and Frederickson, N. (eds) (1996) *Curriculum Related Assessment, Cummins and Bilingual Children.* Clevedon: Multilingual Matters.
Clyne, M. (1980) Triggering and language processing. *Canadian Journal of Psychology* 34 (4), 400–406.
Clyne, M. (1997) Some of the things trilinguals do. *International Journal of Bilingualism* 1 (2), 95–116.
Cohen, A. (1989) Attrition in the productive lexicon of two Portuguese third language speakers. *Studies in Second Language Acquisition* 11, 135–149.
Cohen, A. (1998) *Strategies in Learning and Using a Second Language.* London: Longman.
Cook, V. (1991) The poverty-of-the-stimulus argument and multi-competence. *Second Language Research* 7 (2), 103–117.
Cook, V. (1993a) *Linguistics and Second Language Acquisition.* London: Macmillan.
Cook, V. (1993b) Wholistic multi-competence: jeu d'esprit or paradigm shift? In B. Kettemann and W. Wieden (eds) *Current Issues in European Second Language Acquisition Research* (pp. 3–9). Tübingen: Narr.
Cook, V. (1996) Competence and multi-competence. In G. Brown, K. Malmkjaer and J. Williams (eds) *Performance and Competence in Second Language Acquisition* (pp. 57–69). Cambridge: Cambridge University Press.
Cooper, D. (1999) *Linguistic Attractors: The Cognitive Dynamics of Language Acquisition and Change.* Amsterdam: John Benjamins.
Corder, P. (1971) Idiosyncratic dialects and error analysis. *International Review of Applied Linguistics* 9, 147–159.
Coste, D. (1997) Multilingual and multicultural competence and the role of school. *Language Teaching* 30, 90–93.
Crick, F. (1994) *The Astonishing Hypothesis: The Search for the Soul.* New York: Scribner.
Cummins, J. (1976) The influence of bilingualism on cognitive growth: A synthesis of research findings and explanatory hypotheses. *Working Papers on Bilingualism* 9, 1–43.
Cummins, J. (1979) Cognitive/academic language proficiency, linguistic interdependence, the optimum age question and some other matters. *Working Papers on Bilingualism* 19, 121–129.
Cummins, J. (1984) *Bilingualism and Special Education: Issues in Assessment and Pedagogy.* Clevedon: Multilingual Matters.
Cummins, J. (1991a) Interdependence of first- and second language proficiency. In E. Bialystok (ed.) *Language Processing in Bilingual Children* (pp. 70–89). Cambridge: Cambridge University Press.

Cummins, J. (1991b) Language learning and bilingualism. *Sophia Linguistica* 29, 1–194.
Cummins, J. (2000) Putting language proficiency in its place: Responding to critiques of the conversational/academic distinction. In Cenoz, J. and U. Jessner (eds) *English in Europe: The Acquisition of a Third Language* (pp. 54–83). Clevedon: Multilingual Matters.
de Angelis, G. and Selinker, L. (2001) Interlanguage transfer and competing linguistic systems in the multilingual mind. In J. Cenoz, B. Hufeisen and U. Jessner (eds) *Cross-linguistic Influence in Third Language Acquisition: Psycholinguistic Perspectives* (pp. 42–58). Clevedon: Multilingual Matters.
de Bot, K. (1992) A bilingual production model: Levelt's 'Speaking' model adapted. *Applied Linguistics* 13, 1–24.
de Bot, K. (1996) Language loss. In H. Goebl, P. Nelde, Z. Stáry and W. Wölck (eds) *Kontaktlinguistik. Ein internationales Handbuch zeitgenössischer Forschung. Volume I* (pp. 579–585). Berlin: De Gruyter.
de Bot, K. and Clyne, M. (1989) Language reversion revisited. *Studies in Second Language Acquisition* 11, 167–177.
de Jong, J. and Verhoeven, L. (1992) Modelling and assessing language proficiency. In L. Verhoeven and J. de Jong (eds) *The Construct of Language Proficiency* (pp. 3–19). Amsterdam: John Benjamins.
Dension, N. (1972) Some observations on language variety and plurilingualism. In J. Pride and J. Holmes (eds) *Sociolinguistics* (pp. 65–77). Harmondsworth: Penguin.
Dentler, S., Hufeisen, B. and Lindemann, B. (eds) (2000) *Tertiär- und Drittsprachen: Projekte und empirische Untersuchungen.* Tübingen: Stauffenburg.
Diller, K. (1970) 'Compound' and 'coordinate' bilinguals: A conceptual artifact. *Word* 26, 254–261.
Diller, K. (ed.) (1981) *Individual Differences and Universals in Language Learning Aptitude.* Rowley, MA: Newbury House.
Dörner, D. (1989) *Die Logik des Mißlingens. Strategisches Denken in komplexen Situationen.* Reinbek bei Hamburg: Rowohlt.
Dörnyei, Z. (1998) Motivation in second and foreign language learning. *Language Teaching* 31, 117–135.
Dorian, N. (1973) Grammatical change in a dying dialect. *Language* 49, 413–438.
Dorian, N. (1978) The fate of morphological complexity in language death. *Language* 54, 590–609.
Downes, W. (1984) *Language and Society.* Bungay: Fontana.
Dulay, H., Burt, M. and Krashen, S. (1982) *Language Two.* Oxford: Oxford University Press.
Eco, U. (1994) *The Search for the Perfect Language.* Oxford: Blackwell.
Edwards, J. (1994) *Multilingualism.* London: Routledge.
Eisenstein, M. (1980) Childhood bilingualism and adult language learning aptitude. *International Review of Applied Psychology* 29, 159–174.
Eisenhard, P., Kurth, D. and Stiehl, H. (1995) *Wie Neues entsteht: Die Wissenschaften des Komplexen und Fraktalen.* Reinbek bei Hamburg: Rowohlt.
Ehrman, M. and Oxford, R. (1995) Cognition plus: Correlates of language learning success. *The Modern Language Journal* 79, 67–89.
Ellis, N. (1998) Emergentism, connectionism and language learning. *Language Learning* 48, 631–664.

Ellis, R. (1985) *Understanding Second Language Acquisition*. Oxford: Oxford University Press.
Ellis, R. (1994) *The Study of Second Language Acquisition*. Oxford: Oxford University Press.
Elman, J., Bates, E., Johnson, M., Karmiloff-Smith, A., Parisi, D. and Plunkett, K. (1996) *Rethinking Innateness: A Connectionist Perspective on Development*. Cambridge, MA: MIT Press.
Ervin, S. and Osgood, C. (1954) Second language learning and bilingualism. *Journal of Abnormal and Social Psychology*, Supplement, 49, 139–146.
Faingold, E. (1999) The re-emergence of Spanish and Hebrew in a multilingual adolescent. *International Journal of Bilingual Education and Bilingualism* 2, 283–295.
Fase, W., Jaspaert, K. and Kroon, S. (1992) Maintenance and loss of minority languages: Introductory remarks. In W. Fase, K. Jaspaert and S. Kroon (eds) *Maintenance and Loss of Minority Languages* (pp. 3–13). Amsterdam: John Benjamins.
Feichtinger, A., Lanzmaier-Ugri, K., Farnault, B. and Pornon, J. (2000) Bilder von der Welt in verschiedenen Sprachen. In E. Matzer (ed.) *Sprach- und Kulturerziehung in Theorie und Praxis* (pp. 57–70). Graz: Zentrum für Schulentwicklung, Bereich III: Fremdsprachen.
Felix, S. (1987) *Cognition and Language Growth*. Dordrecht: Foris.
Fetzer, J. (ed.) (1993) *Foundations of Philosophy of Science: Recent Developments*. New York: Paragon House.
Fishman, J. (1972) *The Sociology of Language*. Rowley, MA: Newbury House.
Flege, J. (1998) The role of phonetic category formation in second-language speech learning, In J. Leather and A. James (eds) *New Sounds 97: Proceedings of the Third International Symposium on the Acquisition of Second-language Speech* (pp. 79–88). Klagenfurt: University of Klagenfurt.
Flynn, S. (1987) *A Parameter-setting Model of L2 Acquisition*. Dordrecht: Reidel.
Flynn, S. (1996) A parameter-setting approach to second language acquisition. In W. Ritchie and T. Bhatia (eds) *Handbook of Second Language Acquisition* (pp. 121–159). New York: Academic Press.
Flynn, S. and O'Neil, W. (eds) (1988) *Linguistic Theory and Second Language Acquisition*. Dordrecht: Kluwer.
Fodor, J. (1983) *The Modularity of Mind*. Cambridge, MA: MIT Press.
Franceschini, R. (1999) Sprachadoption: der Einfluss von Minderheitensprachen auf die Mehrheit, oder: Welche Kompetenzen der Minderheitensprachen haben Minderheitensprecher? *Bulletin Suisse de Linguistique Appliquée*, 69 (2), 137–153.
Gardner, H. (1983) *Frames of Mind: Theory of Multiple Intelligences*. New York: Basic Books.
Gardner, R., Lalonde, R., Moorecroft, R. and Evers, F. (1987) Second language attrition: The role of motivation and use. *Journal of Language and Social Psychology* 6, 29–47.
Gardner, R. and MacIntyre, P. (1992) A student's contributions to second-language learning. Part I: Cognitive variables. *Language Teaching* 25, 211–220.
Gardner, R. and MacIntyre, P. (1993) A student's contributions to second-language learning. Part II: Affective variables. *Language Teaching* 26, 1–11.
Gardner, R. and MacIntyre, P. (1994) The subtle effects of language anxiety on cognitive processing in the second language. *Language Learning* 44 (2), 283–305.
Gardner, R., Tremblay, P. and Masgoret, A-M. (1997) Towards a full model of

second language learning: An empirical investigation. *The Modern Language Journal* 81, 344–362.

Gardner-Chloros, P. (1995) Code-switching in community, regional and national repertoires: The myth of the discreteness of linguistic systems. In L. Milroy and P. Muysken (eds) *One Speaker, Two Languages: Cross-disciplinary Perspectives on Code-switching* (pp. 68–89). Cambridge: Cambridge University Press.

Gass, S. and Selinker, L. (1994) *Second Language Acquisition: An Introductory Course*. Hillsdale, NJ: Lawrence Erlbaum.

Gatto, D. (2000) Language proficiency and narrative proficiency of a trilingual child. In S. Dentler, B. Hufeisen and B. Lindemann (eds) *Tertiär- und Drittsprachen: Projekte und empirische Untersuchungen* (pp. 117–142). Tübingen: Stauffenburg.

Genesee, F., Tucker, R. and Lambert, W. (1975) Communication skills in bilingual children. *Child Development* 54, 105–114.

Gigerenzer, G. and Murray, J. (1989) *Cognition as Intuitive Statistics*. Hillsdale, NJ: Lawrence Erlbaum.

Gillette, B. (1987) Two successful language learners: An introspective approach. In C. Faerch and G. Kasper (eds) *Introspection in Second Language Research* (pp. 269–279). Clevedon: Multilingual Matters.

Gleick, J. (1987) *Chaos: Making a New Science*. New York: Viking.

Gombert, E. (1992) *Metalinguistic Development*. New York: Harvester Wheatsheaf.

Gottman, J. (1995) *The Analysis of Change*. Hillsdale, NJ: Lawrence Erlbaum.

Gregg, K. (1989) Second language acquisition theory: The case for a generative perspective. In S. Gass and J. Schachter (eds) *Linguistic Perspectives on Second Language Acquisition* (pp. 15–40). Cambridge: Cambridge University Press.

Gregg, K. (1990) The variable competence model of second language acquisition and why it isn't. *Applied Linguistics* 14, 364–383.

Green, D. (1986) Control, activation and resource: A framework and a model for the control of speech in bilinguals. *Brain and Language* 27, 210–223.

Grosjean, F. (1982) *Life with Two Languages*. Cambridge, MA: Harvard University Press.

Grosjean, F. (1985) The bilingual as a competent but specific speaker-hearer. *Journal of Multilingual and Multicultural Development* 6, 467–477.

Grosjean, F. (1992) Another view of bilingualism. In R. Harris (ed.) *Cognitive Processing in Bilinguals* (pp. 51–62). Amsterdam: North Holland.

Grosjean, F. (1998) Studying bilinguals: Methodological and conceptual issues. *Bilingualism: Language and Cognition* 1, 131–149.

Grosjean, F. (2001) The bilingual's language modes. In J. Nicol (ed.) *One Mind, Two Languages: Bilingual Language Processing* (pp. 1–25). Oxford: Blackwell.

Grosjean, F. and Py, B. (1991) La restructuration d'une première langue: l'intégration de variantes de contact dans la compétence de migrants bilingues. *La Linguistique* 27, 35–60.

Gumperz, J. (1972) Sociolinguistics and communication in small groups. In J. Pride and J. Holmes (eds) *Sociolinguistics: Selected Readings* (pp. 203–224). Harmondsworth: Penguin.

Haarmann, H. (1980) *Multilingualismus (1). Probleme der Systematik und Typologie*. Tübingen: Narr.

Haken, H. (1981) *Erfolgsgeheimnisse der Natur. Synergetik: Die Lehre vom Zusammenwirken*. Stuttgart: Deutsche Verlagsanstalt.

Hakuta, K. (1986) *Mirror of Language: The Debate on Bilingualism*. New York: Basic Books.
Hamers, J. and Blanc, M. (1989) *Bilinguality and Bilingualism*. Cambridge: Cambridge University Press.
Hansegård, N. (1975) Tvåspråkighet eller halvspråkighet? *Aldus* Series 253, Stockholm.
Harley, B. (1994) Maintaining French as a second language in adulthood. *The Canadian Modern Language Review* 50 (4), 688–713.
Harley, B., Allen, P., Cummins, J. and Swain, M. (eds) (1990a) *The Development of Second Language Proficiency*. Cambridge: Cambridge University Press.
Harley, B., Cummins, J., Swain, M. and Allen, P. (1990b) The nature of language proficiency. In B. Harley, P. Allen, J. Cummins and M. Swain (eds) *The Development of Second Language Proficiency* (pp. 7–25). Cambridge: Cambridge University Press.
Harley, B. and Hart, D. (1997) Language aptitude and second language proficiency in classroom learners of different starting ages. *Studies in Second Language Acquisition* 19, 379–400.
Haugen, E. (1969) *The Norwegian Language in America: A Study in Bilingual Behavior*. Bloomington: Indiana University Press.
Haugen, E. (1972) The stigmata of bilingualism. In A. Dil (ed.) *The Ecology of Language* (pp. 307–324). Stanford: Stanford University Press.
Heller, M. (ed.) (1988) *Codeswitching: Anthropological and Sociolinguistic Perspectives*. The Hague: Mouton de Gruyter.
Herdina, P. (1988) Menschenautomat und Automatenmensch. Eine philosophische Reflexion. In H. Czuma (ed.) *Menschenbilder* (pp. 129–154). Wien: Verlag für Gesellschaftskritik.
Herdina, P. (1990) *Methodologie und Krise. Skizze eines metatheoretischen Paradigmenwechsels*. Innsbruck: Institut für Sprachwissenschaft.
Herdina, P. (ed.) (1992a) *Methodenfragen der Geisteswissenschaften*. Innsbruck: Institut für Sprachwissenschaft.
Herdina, P. (1992b) Die Bewertbarkeit linguistischer Theorien. Unterdeterminiertheit in sprachwissenschaftlicher Theoriebildung. In P. Herdina (ed.) *Methodenfragen der Geisteswissenschaften* (pp. 167–204). Innsbruck: Institut für Sprachwissenschaft.
Herdina, P. (1996) Structures, transformations and notions. In P. Herdina, U. Jessner and M. Kienpointner (eds) *Language Acquisition and Syntactic Structures* (pp. 35–71). Innsbruck: Institut für Sprachwissenschaft.
Herdina, P. and U. Jessner (1998) Language maintenance in multilinguals. A psycholinguistic perspective. In J. Leather and A. James (eds) *New Sounds 97: Proceedings of the Third International Symposium on the Acquisition of Second-language Speech* (pp. 135–143). Klagenfurt: University of Klagenfurt.
Hoffmann, C. (1985) Language acquisition in two trilingual children. *Journal of Multilingual and Multicultural Development* 6, 281–287.
Hoffmann, C. (1991) *An Introduction to Bilingualism*. London: Longman.
Hoffmann, C. and Widdicombe, S. (1998) The language behaviour of trilingual children: developmental aspects. Paper held at EUROSLA 8 Conference in Paris, September.
Huebner, T. (1991) Second language acquisition: Litmus test for linguistic theory?

In T. Huebner and C. Ferguson (eds) *Crosscurrents in Second Language Acquisition* (pp. 3–22). Amsterdam: John Benjamins.

Hufeisen, B. (2000a) How do foreign language learners evaluate various aspects of their multilingualism? In S. Dentler, B. Hufeisen and B. Lindemann (eds) *Tertiär- und Drittsprachen: Projekte und empirische Untersuchungen* (pp. 23–29). Tübingen: Stauffenburg.

Hufeisen, B. (2000b) A European perspective: Tertiary languages with a focus on German as L3 (pp. 209–229). In J. Rosenthal (ed.) *Handbook of Undergraduate Second Language Education*. Mahwah, NJ: Lawrence Erlbaum.

Hufeisen, B. and Lindemann, B. (eds) (1998) *L2–L3 und ihre zwischensprachliche Interaktion: Zu individueller Mehrsprachigkeit und gesteuertem Lernen*. Tübingen: Stauffenburg.

Hyltenstam, K. and Obler, L. (eds) (1989) *Bilingualism Across the Lifespan: Aspects of Acquisition, Maturity and Loss*. Cambridge: Cambridge University Press.

Hyltenstam, K. and Stroud, C. (1993) Second language regression in Alzheimer's dementia. In K. Hyltenstam and A. Viberg (eds) *Progression and Regression in Language: Sociocultural, Neuropsychological and Linguistic Perspectives* (pp. 222–242). New York: Academic Press.

Hyltenstam, K. and Stroud, C. (1996) Language maintenance. In H. Goebl, P. Nelde, Z. Stáry and W. Wölck (eds) *Kontaktlinguistik. Ein internationales Handbuch zeitgenössischer Forschung. Vol.1* (pp. 567–578). Berlin: De Gruyter.

Hyltenstam, K. and Viberg, A. (eds) (1993) *Progression and Regression in Language: Sociocultural, Neuropsychological and Linguistic Perspectives*. New York: Academic Press.

Hymes, D. (1972) On communicative competence. In J. Pride and J. Holmes (eds) *Sociolinguistics. Selected Readings* (pp. 269–294). Harmondsworth: Penguin.

Illitch, I. and Sanders, B. (1988) *ABC: Alphabetisation of the Popular Mind*. Berkeley, CA: North Point Press.

Jakobovits, L. (1969) Second language learning and transfer theory: A theoretical assessment. *Language Learning* 19, 55–86.

James, C. (1992) Awareness, consciousness and language contrast. In C. Mair and M. Markus (eds) *New Departures in Contrastive Linguistics*. Vol. 2 (pp. 183–198). Innsbruck: Institut für Anglistik.

James, C. (1998) *Errors in Language Learning and Use*. London: Longman.

Jessner, U. (1995) How beneficial is bilingualism? Cognitive aspects of bilingual proficiency. In K. Sornig, D. Halwachs, C. Penzinger and G. Ambrosch (eds) *Linguistics With a Human Face* (pp. 173–182). Graz: Institut für Sprachwissenschaft.

Jessner, U. (1999) Metalinguistic awareness in multilinguals: Cognitive aspects of third language learning. *Language Awareness* 8 (3&4), 201–209.

Jessner, U. (2001) Drittspracherwerb: Implikationen für einen Sprachenunterricht der Zukunft. In S. Kuri and R. Saxer (eds) *Deutsch als Fremdsprache an der Schwelle zum 21. Jahrhundert: Zukunftsorientierte Konzepte und Projekte* (pp. 54–64). Innsbruck, Wien: Studienverlag.

Jessner, U. and Herdina, P. (1994) The paradox of transfer. Paper held at IRAAL Conference in Dublin, June.

Jessner, U. and Herdina, P. (1996) Interaktionsphänomene im multilingualen Menschen: Erklärungsmöglichkeiten durch einen systemtheoretischen Ansatz.

In A. Fill (ed.) *Sprachökologie und Ökolinguistik* (pp. 217–230). Tübingen: Stauffenburg.
Jespersen, O. (1922) *Language*. London: Allen and Unwin.
Jisa, H. (1999) Some dynamics of bilingual development. *Acquisition et Interaction en Langue Etrangère* 1, 7–32.
Kachru, Y. (1994) Monolingual bias in SLA research. *TESOL Quarterly* 28, 797–800.
Karmiloff-Smith, A. (1992) *Beyond Modularity: A Developmental Perspective on Cognitive Science*. Cambridge, MA: MIT Press.
Karpf, A. (1990) *Selbstorganisationsprozesse in der sprachlichen Ontogenese: Erst- und Fremdsprache(n)*. Tübingen: Narr.
Kasher, A. (ed.) (1991) *The Chomskyan Turn*. London: Blackwell.
Kasper, G. (1992) Pragmatic transfer. *Second Language Research* 8, 203–231.
Kasper, G. and Blum-Kulka, S. (eds) (1993) *Interlanguage Pragmatics*. Oxford: Oxford University Press.
Kasper, G. and Kellerman, E. (eds) (1999) *Communication Strategies: Psycholinguistic and Sociolinguistic Perspectives*. London: Longman.
Keatley, C. (1992) History of bilingualism research in cognitive psychology. In R. Harris (ed.) *Cognitive Processing in Bilinguals* (pp. 15–49). Amsterdam: North Holland.
Kecskes, I. and Papp, T. (2000) *Foreign Language and Mother Tongue*. Mahwah, NJ: Lawrence Erlbaum.
Kellerman, E. and Sharwood Smith, M. (eds) (1986) *Crosslinguistic Influence in Second Language Acquisition*. Oxford: Pergamon Press.
Kellerman, E. (1995) Crosslinguistic influence: Transfer to Nowhere? *Annual Review of Applied Linguistics* 15, 125–150.
Kelly, L.G. (ed.) (1969) *Description and Measurement of Bilingualism: An International Seminar at the University of Moncton, June 6–14, 1967*. Toronto: University of Toronto Press.
Kim, K., Relkin, N., Kyoung-Min, L. and Hirsch, J. (1997) Distinct cortical areas associated with native and second languages. *Nature* 388, 171–174.
Klein, E. (1995) Second vs. third language acquisition: Is there a difference? *Language Learning* 45 (3), 419–465.
Klein, W. (1986) *Second Language Acquisition*. Cambridge: Cambridge University Press.
Klein, W. (1991) SLA theory: Prolegomena to a theory of language acquisition and implications for theoretical linguistics. In T. Huebner and C. Ferguson (eds) *Crosscurrents in Second Language Acquisition* (pp. 169–194). Amsterdam: John Benjamins.
Kramarae, C. (1982) Gender: How she speaks. In H. Giles and E. Bouchard (eds) *Attitudes Towards Language Variation: Social and Applied Contexts* (pp. 84–115). London: Edward Arnold.
Krashen, S. (1981) *Second Language Acquisition and Second Language Learning*. Oxford: Pergamon Press.
Krashen, S. (1985) *The Input Hypothesis: Issues and Implications*. New York: Longman.
Krohn, W. and Küppers, G. (eds) (1992) *Emergenz: Die Entstehung von Ordnung, Organisation und Bedeutung*. Frankfurt: Suhrkamp.
Kuhs, K. (1989) *Sozialpsychologische Faktoren im Zweitspracherwerb: eine Untersuchung bei griechischen Migrantenkindern in der Bundesrepublik Deutschland*. Tübingen: Narr.

Labov, W. (1969) The logic of non-standard English. *Georgetown Monographs in Languages and Linguistics* 22, 1–45.
Lafont, R. (1990) Codeswitching et production du sens. In R. Jacobson (ed.) *Codeswitching as a Worldwide Phenomenon* (pp. 71–85). New York: Lang.
Lambert, W. (1977) The effects of bilingualism on the individual: cognitive and sociocultural consequences. In P. Hornby (ed.) *Bilingualism: Psychological, Social and Educational Implications* (pp. 15–28). New York: Academic Press.
Landry, R., Allard, R. and Théberge, R. (1991) School and family French ambiance and the bilingual development of francophone Western Canadians. *Canadian Modern Language Review* 47, 878–915.
Larsen-Freeman, D. (1997) Chaos/complexity science and second language acquisition. *Applied Linguistics* 18, 141–165.
Larsen-Freeman, D. and Long, M. (1991) *An Introduction to Second Language Acquisition Research*. London: Longman.
Lasagabaster, D. (1997) Creatividad y concienca metalingüística: incidencia en el apprendizaje del inglés como L3. Unpublished doctoral thesis, University of the Basque Country, Vitoria-Gasteiz.
Lenneberg, E. (1967) *Biological Foundations of Language*. New York: Wiley.
Leopold, F. (1939–1949) *Speech Development of a Bilingual Child: A Linguist's Record*. Vol. 1–4. Evanston, IL: Northwestern University Press.
Lightfoot, D. (1999) *The Development of Language: Acquisition, Change and Evolution*. London: Blackwell.
Lüdi, G. (1996) Mehrsprachigkeit. In H. Goebl, P. Nelde, Z. Starý and W. Wölck (eds) *Kontaktlinguistik: ein internationales Handbuch zeitgenössischer Forschung* (pp. 233–245). Berlin: Walter de Gruyter.
Lurija, A. (1991) *Der Mann, dessen Welt in Scherben ging. Zwei neurologische Geschichten*. Reinbek bei Hamburg: Rowohlt.
Lyons, J. (1996) On competence and performance and related notions. In G. Brown, K. Malmkjaer and J. Williams, J. (eds) *Performance and Competence in Second Language Acquisition* (pp. 11–32). Cambridge: Cambridge University Press.
MacIntyre, P. and Gardner, R. (1989) Anxiety and second-language learning: Toward a theoretical clarification. *Language Learning* 39 (2), 251–275.
MacIntyre, P. and Gardner, R. (1991a) Methods and results in the study of anxiety and language learning: A review of the literature. *Language Learning* 41 (1), 85–117.
MacIntyre, P. and Gardner, R. (1991b) Language anxiety: Its relationship to other anxieties and to processing in native and second language. *Language Learning* 41 (4), 513–534.
Mackey, W. (1969) Introduction: How can bilingualism be described and measured? Outline of the problems. In L.G. Kelly (ed.) *Description and Measurement of Bilingualism: An International Seminar at the University of Moncton, June 6–14, 1967* (pp. 1–9). Toronto: University of Toronto Press.
MacWhinney, B. (1992) Competition and transfer in second language learning. In R. Harris (ed.) *Cognitive Processing in Bilinguals* (pp. 371–390). Amsterdam: North Holland.
MacWhinney, B. (ed.) (1999) *The Emergence of Language*. Mahwah, NJ: Lawrence Erlbaum.
MacWhinney, B. and Bates, E. (1989) (eds) *The Crosslinguistic Study of Sentence Processing*. Cambridge: Cambridge University Press.

Malakoff, M. (1992) Translation skill and metalinguistic awareness in bilinguals. In R. Harris (ed.) *Cognitive Processing in Bilinguals* (pp. 515–530). Amsterdam: North Holland.

Maturana, H. (1998) *Biologie der Realität*. Frankfurt: Suhrkamp.

Maturana, H. and Varela, F. (1987) *Der Baum der Erkenntnis. Die biologischen Wurzeln des menschlichen Erkennens*. München: Goldmann.

Mißler, B. (2000) Previous experience of foreign language learning and its contribution to the development of learning strategies. In S. Dentler, B. Hufeisen and B. Lindemann (eds) *Tertiär- und Drittsprachen: Projekte und empirische Untersuchungen* (pp. 7–21). Tübingen: Stauffenburg.

McDonough, S. (1999) Learner strategies. *Language Teaching* 32, 1–18.

McLaughlin, B. (1987) *Theories of Second-language Learning*. London: Edward Arnold.

McLaughlin, B. (1990) The relationship between first and second languages: Language proficiency and language aptitude. In B. Harley, P. Allen, J. Cummins and M. Swain (eds) *The Development of Second Language Proficiency* (pp. 158–174). Cambridge: Cambridge University Press.

McLaughlin, B. and Heredia, R. (1996) Information-processing approaches to research on second language acquisition and use. In W. Ritchie and T.K. Bhatia (eds) *Handbook of Second Language Acquisition* (pp. 213–228). San Diego: Academic Press.

McLaughlin, B. and Nayak, N. (1989) Processing a new language: Does knowing other languages make a difference? In H. Dechert and M. Raupach (eds) *Interlingual Processes* (pp. 5–14). Tübingen: Narr.

McMahon, A. (1994) *Understanding Language Change*. Cambridge: Cambridge University Press.

Meara, P. (1997) Towards a new approach to modelling vocabulary acquisition. In N. Schmitt and M. McCarthy (eds) *Vocabulary: Description, Acquisition and Pedagogy* (pp. 109–121). Cambridge: Cambridge University Press.

Meara, P. (1999) Self organization in bilingual lexicons. In P. Broeder and J. Muure (eds) *Language and Thought in Development* (pp. 127–144). Tübingen: Narr.

Milroy, L. and Muysken, P. (eds) (1995) *One Speaker, Two Languages: Cross-disciplinary Perspectives on Code-switching*. Cambridge: Cambridge University Press.

Multhaup, U. (1997) Mental networks, procedural knowledge and foreign language teaching. *Language Awareness* 6 (2&3), 75–92.

Myers Scotton, C. (1990) Codeswitching and borrowing: Interpersonal and macrolevel meaning. In R. Jacobson (ed.) *Codeswitching as a Worldwide Phenomenon* (pp. 85–111). New York: Lang.

Nagel, W. (2000) Zwei Beispiele für den fächerübergreifenden Lateinunterricht. In E. Matzer (ed.) *Sprach- und Kulturerziehung in Theorie und Praxis* (pp. 70–7). Graz: Zentrum für Schulentwicklung, Bereich III: Fremdsprachen.

Naiman, N., Fröhlich, M., Stern, H. and Todesco, A. (1996 [1978]) *The Good Language Learner*. Clevedon: Multilingual Matters.

Nation, R. and McLaughlin, B. (1986) Experts and novices: An information-processing approach to the 'good language learner' problem. *Applied Psycholinguistics* 7, 51–56.

Nemser, W. (1971) Approximative systems of foreign language learners. *International Review of Applied Linguistics* 9, 115–123.

Neufeld, G. (1979) Towards a theory of language learning ability. *Language Learning* 29 (2), 227–241.
Nicolis, G. and Prigogine, I. (1989) *Exploring Complexity: An Introduction*. New York: W.H. Freeman.
Nørretranders, T. (1997) *Spüre die Welt. Die Wissenschaft des Bewußtseins*. Reinbek bei Hamburg: Rowohlt.
North, B. (1997) Perspectives on language proficiency and aspects of competence. *Language Teaching* 30, 93–100.
Oakes, G. (1994) Ideal types. In R. Asher (ed.) *The Encyclopedia of Language and Linguistics* (p. 1638). Oxford: Pergamon.
Odlin, T. (1989) *Language Transfer: Cross-linguistic Influence in Language Learning*. Cambridge: Cambridge University Press.
Oksaar, E. (1990) Language contact and culture contact: Towards an integrative approach in second language acquisition research. In H. Dechert (ed.) *Current Trends in European Second Language Acquisition Research* (pp. 230–243). Clevedon: Multilingual Matters.
O'Malley, J. and Chamot, A-U. (1990) *Learning Strategies in Second Language Learning*. Cambridge: Cambridge University Press.
Oller, J. (1976) Evidence for a general language proficiency factor: An expectancy grammar. *Die Neueren Sprachen* 75, 165–174.
Ornstein, R. and Thompson, R. (1984) *The Amazing Brain*. Boston, MA: Houghton Mifflin.
Osgood, C. (1953) *Method and Theory in Experimental Psychology*. New York: Oxford University Press.
Oxford, R. (1990a) *Language Learning Strategies: What Every Teacher Should Know*. New York: Harper Collins.
Oxford, R. (1990b) Styles, strategies, and aptitude: Connections for language learning. In T. Parry and C. Stansfield (eds) *Language Aptitude Reconsidered* (pp. 67–125). Englewood Cliffs, NJ: Prentice Hall.
Paivio, A. (1986) *Mental Representations: A Dual Coding Approach*. Oxford: Oxford University Press.
Palmen, M-J., Bongaerts, T. and Schils, E. (1997) L'authenticité de la prononciation dans l'acquisition d'une langue étrangère au-dela de la période critique: des apprenants neérlandais parvenus à un niveau très avancé en français. *Acquisition et Interaction en Langue Etrangère* 9, 173–191.
Pan, B. and Berko Gleason, J. (1986) The study of language loss: Models and hypotheses for an emerging discipline. *Applied Psycholinguistics* 7, 193–206.
Pandharipande, R. (1990) Formal and functional constraints on code-mixing. In R. Jacobson (ed.) *Codeswitching as a Worldwide Phenomenon* (pp. 15–32). New York: Lang.
Parry, T. and Stansfield, C. (eds) (1990) *Language Aptitude Reconsidered*. Englewood Cliffs, NJ: Prentice Hall.
Peal, E. and Lambert, W. (1962) The relation of bilingualism to intelligence. *Psychological Monographs* 76, 1–23.
Peltzer-Karpf, A. (1996) Early foreign language learning: A biological perspective. In B. Kettemann and I. Landsiedler (eds) *The Effectiveness of Language Learning and Teaching* (pp. 90–105). Graz: Zentrum für Schulentwicklung.
Penfield, W. and Roberts, L. (1959) *Speech and Brain Mechanisms*. Princeton, NJ: Princeton University Press.

Perecman, E. (1984) Spontaneous transmission and language mixing in a polyglot aphasic. *Brain and Language* 23, 43–63.
Pfaff, C. (1979) Constraints on language mixing. *Language* 55 (2), 291–318.
Phillips, D. (1992) *The Social Scientist's Bestiary: A Guide to Fabled Threats to, and Defences of, Naturalistic Social Science*. Oxford: Pergamon Press.
Piatelli-Palmarini, M. (1980) *Language and Learning: The Debate between Jean Piaget and Noam Chomsky*. Cambridge, MA: Harvard University Press.
Pinker, S. (1984) *Language Learnability and Language Development*. Cambridge, MA: Harvard University Press.
Pinker, S. (1989) *Learnability and Cognition: The Acquisition of Argument Structure*. Cambridge, MA: MIT Press.
Poplack, S. (1980) 'Sometimes I'll start a sentence in English y termino en espagnol' : Toward a typology of code-switching. *Linguistics* 18, 581–618.
Poulisse, N. (1997) Language production in bilinguals. In A. de Groot and J. Kroll (eds) *Tutorials in Bilingualism: Psycholinguistic Perspective* (pp. 201–224). Mahwah, NJ: Lawrence Erlbaum.
Prigogine, I. (1980) *From Being to Becoming*. San Francisco: W.H. Freeman.
Reynolds, A. (ed.) (1991a) *Bilingualism, Multiculturalism, and Second Language Learning*. Hillsdale, NJ: Lawrence Erlbaum.
Reynolds, A. (1991b) The cognitive consequences of bilingualism. In A. Reynolds (ed.) *Bilingualism, Multiculturalism, and Second Language Learning* (pp. 145–182). Hillsdale, NJ: Lawrence Erlbaum.
Ricciardelli, L. (1992) Bilingualism and cognitive development in relation to Threshold Theory. *Journal of Psycholinguistic Research* 21, 56–67.
Richards, J. (1974) Social factors, interlanguage and language learning. In J. Richards (ed.) *Error Analysis: Perspectives on Second Language Acquisition* (pp. 64–91). London: Longman.
Ridley, J. and Singleton, D. (1995) Contrastivity and individual learner contrasts. *Fremdsprachen Lernen und Lehren* 24, 123–137.
Ringbom, H. (1987) *The Role of the First Language in Foreign Language Learning*. Clevedon: Multilingual Matters.
Robert, J-M. (1994) *L'Aventure des Neurones*. Paris: Editions du Seuil.
Robertson, S., Cohen, A. and Mayer-Kress, G. (1993) Behavioral chaos: Beyond the metaphor. In L. Smith and E. Thelen (eds) *A Dynamic Systems Approach to Development: Applications* (pp. 120–150). Cambridge, MA: MIT Press.
Roth, G. (1994) *Das Gehirn und seine Wirklichkeit. Kognitive Neurobiologie und ihre philosophischen Konsequenzen*. Frankfurt: Suhrkamp.
Romaine, S. (1989) *Bilingualism*. London: Blackwell.
Ronjat, J. (1913) *Le développement du langage observé chez un enfant bilingue*. Paris: Champion.
Rumelhart, D. and McClelland, J. (1986) *Parallel Distributed Processing: Explorations in the Microstructure of Cognition*. Cambridge, MA: MIT Press.
Ryle, G. (1973 [1948]) *The Concept of Mind*. Harmondsworth: Penguin.
Saer, D. (1922) An inquiry into the effect of bilingualism upon the intelligence of young children. *Journal of Experimental Pedagogy* 6, 232–240 & 266–274.
Saer, D. (1923) The effects of bilingualism on intelligence. *British Journal of Psychology* 14, 25–38.
Salkie, R. (1990) *The Chomsky Update: Linguistics and Politics*. London: Routledge.

Schumann, J. (1978) *The Pidginization Process: A Model for Second Language Acquisition*. Rowley, MA: Newbury House.

Schumann, J. (1986) Research on the acculturation model for second language acquisition. *Journal of Multilingual and Multicultural Development* 7, 379–392.

Schumann, J. (1990) Extending the scope of the acculturation/pidginization model to include cognition. *TESOL Quarterly* 24 (4), 667–684.

Schumann, J. (1997) *The Neurobiology of Affect in Language*. Oxford: Blackwell.

Schweizer, H. (1979) *Sprache und Systemtheorie*. Tübingen: Narr.

Seliger, H. and Vago, R. (eds) (1991) *First Language Attrition*. Cambridge: Cambridge University Press.

Seliger, H. (1996) Primary language attrition in the context of bilingualism. In W. Ritchie and T. Bathia (eds) *Handbook of Second Language Acquisition* (pp. 605–627). New York: Academic Press.

Selinker, L. (1972) Interlanguage. *International Review of Applied Linguistics* 10 (3), 209–231.

Selinker, L. (1974) Interlanguage. In J. Richards (ed.) *Error Analysis: Perspectives on Second Language Acquisition* (pp. 31–54). London: Longman.

Selinker, L. (1992) *Rediscovering Interlanguage*. London: Longman.

Selinker, L. (1996) On the notion of 'IL competence' in early SLA research: An aid to understanding some baffling current issues. In G. Brown, K. Malmkjaer and J. Williams (eds) *Performance and Competence in Second Language Acquisition* (pp. 89–113). Cambridge: Cambridge University Press.

Shannon, B. (1991) Faulty language selection in polyglots. *Language and Cognitive Processes* 6 (4), 339–350.

Sharwood Smith, M. (1989) Crosslinguistic influence in language loss. In K. Hyltenstam and L. Obler (eds) *Bilingualism Across the Lifespan: Aspects of Acquisition, Maturity and Loss* (pp. 185–201). Cambridge: Cambridge University Press.

Sharwood Smith, M. (1994) *Second Language Learning*. London: Longman.

Sharwood Smith, M. and Kellerman, E. (1989) The interpretation of second language output. In H. Dechert and M. Raupach (eds) *Transfer in Language Productions* (pp. 217–235). Norwood, NJ: Ablex.

Shore, C. (1995) *Individual Differences in Language Development*. Thousand Oaks: Sage Publications.

Simon, F. (1997) *Lebende Systeme: Wirklichkeitskonstruktionen in der systemischen Therapie*. Frankfurt/Main: Suhrkamp.

Singleton, D. (1989) *Language Acquisition: The Age Factor*. Clevedon: Multilingual Matters.

Singleton, D. (1995) Introduction: A critical look at the critical period hypothesis in second language acquisition research. In D. Singleton and Z. Lengyel (eds) *The Age Factor in Second Language Acquisition* (pp. 1–29). Clevedon: Multilingual Matters.

Singleton, D. (1996) Crosslinguistic lexical operations and the L2 mental lexicon. In T. Hickey and J. Williams (eds) *Language, Education and Society in a Changing World* (pp. 246–252). Clevedon: Multilingual Matters.

Singleton, D. and Lengyel, Z. (eds) (1995) *The Age Factor in Second Language Acquisition*. Clevedon: Multilingual Matters.

Skehan, P. (1989) *Individual Differences in Second-language Learning*. London: Edward Arnold.

Skehan, P. (1998) *A Cognitive Approach to Language Learning* Oxford: Oxford University Press.
Skutnabb-Kangas, T. (1976) Bilingualism, semilingualism and school achievement. *Linguistische Berichte* 45, 55–64.
Skutnabb-Kangas, T. (1984) *Bilingualism or Not: The Education of Minorities.* Clevedon: Multilingual Matters.
Smith, L. and Thelen, E. (eds) (1993) *A Dynamic Systems Approach to Development: Applications.* Cambridge, MA: MIT Press.
Smith, N. and Tsimpli, I-M. (1995) *The Mind of a Savant.* London: Blackwell.
Sorace, A. (1998) Residual L2 optionality and L1 attrition. Paper held at EUROSLA 8 conference in Paris, September.
Sorace, A., Heycock, C. and Shillcock, R. (1998) Introduction: Trends and convergences in language acquisition research. *Lingua* 106, 1–21.
Springer, S. and Deutsch, G. (1981) *Left Brain, Right Brain: Perspectives from Cognitive Neuroscience.* New York: W.H. Freeman and Company.
Sridhar, S. (1994) A reality check for SLA theories. *TESOL Quarterly* 28, 800–804.
Sridhar, S. and Sridhar, K. (1980) The syntax and psycholinguistics of bilingual codemixing. *Canadian Journal of Psychology* 34 (4), 407–416.
Stadler, M. and Kruse, P. (1992) Zur Emergenz psychischer Qualitäten. In W. Krohn and G. Küppers (eds) *Emergenz: Die Entstehung von Ordnung, Organisation und Bedeutung* (pp. 134–160). Frankfurt: Suhrkamp.
Sternberg, J. (1988) *The Triarchic Mind.* New York: Viking.
Stork, I. (1996) Die Rolle des Ökonomiebegriffs in der Ökolinguistik. In A. Fill (ed.) *Sprachökologie und Ökolinguistik* (pp. 93–102). Tübingen: Stauffenburg.
Strohner, H. (1995) *Kognitive Systeme. Eine Einführung in die Kognitionswissenschaft.* Opladen: Westdeutscher Verlag.
Tarone, E. (1983) On the variability of interlanguage systems. *Applied Linguistics* 4, 143–163.
Tarone, E. (1988) *Variation in Interlanguage.* London: Edward Arnold.
Timm, L. (1975) Spanish–English code-switching: el porque y how-not-to. *Romance Philology* 28, 437–482.
Thomas, J. (1988) The role played by metalinguistic awareness in second and third language learning. *Journal of Multilingual and Multicultural Development* 9, 235–246.
Titone, R. (1994) Bilingual education and the development of metalinguistic abilities: A research project. *International Journal of Psycholinguistics* 10 (1), 5–14.
Towell, R. and Hawkins, R. (1994) *Approaches to Second Language Acquisition.* Clevedon: Multilingual Matters.
Tucker, M. and Hirsh-Pasek, K. (1993) Systems and language: Implications for acquisition. In L. Smith and E. Thelen (eds) (1993) *A Dynamic Systems Approach to Development: Applications* (pp. 359–384). Cambridge, MA: MIT Press.
Tunmer, W., Pratt, C. and Herriman, M. (eds) (1984) *Metalinguistic Awareness in Children.* Berlin: Springer.
Turian, D. and Altenberg, E. (1991) Compensatory strategies of child first language attrition. In H. Seliger and R. Vago (eds) *First Language Attrition* (pp. 207–226). Cambridge: Cambridge University Press.
Uexkuell, J. (1973) *Theoretische Biologie.* Frankfurt am Main: Suhrkamp.
Ushioda, E. (1996) Developing a dynamic concept of L2 motivation. In T. Hickey

and J. Williams (eds) *Language, Education and Society in a Changing World* (pp. 239–245). Clevedon: Multilingual Matters.
Vaid, J. and Hall, G. (1991) Neuropsychological perspectives on bilingualism: Right, left and center. In A. Reynolds (ed.) *Bilingualism, Multiculturalism, and Second Language Learning* (pp. 81–112). Hillsdale, NJ: Lawrence Erlbaum.
van Geert, P. (1993) A dynamic systems model of cognitive growth: Competition and support under limited resource conditions. In L. Smith and E. Thelen (eds) (1993) *A Dynamic Systems Approach to Development: Applications* (pp. 265–331). Cambridge, MA: MIT Press.
van Geert, P. (1994) *Dynamic Systems of Development: Change Between Complexity and Chaos*. New York: Harvester Wheatsheaf.
van Kleeck, A. (1982) The emergence of linguistic awareness: A cognitive framework. *Merrill-Palmer Quarterly* 28, 237–265.
Verhoeven, L. (1992) Assessment of bilingual proficiency. In L. Verhoeven and J. de Jong (eds) *The Construct of Language Proficiency* (pp. 125–136). Amsterdam: John Benjamins.
Verhoeven, L. (1994) Transfer in bilingual development: The linguistic interdependence hypothesis revisited. *Language Learning* 44, 381–415.
Vincent, J-D. (1986) *Biologie des Passions*. Paris: Editions Odile Jacob.
Vogel, K. (1990) *Lernersprache: linguistische und psycholinguistische Grundfragen zu ihrer Erforschung*. Tübingen: Narr.
Vollmer, H. (1982) *Spracherwerb und Sprachbeherrschung. Untersuchungen zur Struktur von Fremdsprachenfähigkeit*. Tübingen: Narr.
von Bertalanffy, L. (1968) *General System Theory: Foundation, Developments, Applications*. New York: Braziller.
Waddington, C. (1977) *Tools for Thought*. Fontana: Bungay.
Waldorp, M. (1992) *Complexity: The Emerging Science at the Edge of Order and Chaos*. New York: Simon and Schuster.
Wardhaugh, R. (1993) *Investigating Language: Central Problems in Linguistics*. Oxford: Blackwell.
Weinreich, U. (1953) *Languages in Contact: Findings and Problems*. The Hague: Mouton.
Weisgerber, L. (1966) Vorurteile und Gefahren der Zweisprachigkeit. *Wirkendes Wort* 16, 73–89.
Weltens, B. (1988) *The Attrition of French as a Foreign Language*. Dordrecht: Foris.
Weltens, B., de Bot, K. and van Els, T. (eds) (1986) *Language Attrition in Progress*. Dordrecht: Foris.
Weltens, B. and Grendel, M. (1993) Attrition of vocabulary knowledge. In R. Schreuder and B. Weltens (eds) *The Bilingual Lexicon* (pp. 135–156). Amsterdam: John Benjamins.
Wessels, M. (1982) *Cognitive Psychology*. New York: Harper and Row.
White, L. (1989) *Universal Grammar and Second Language Acquisition*. Amsterdam: John Benjamins.
White, L. (1998) The implications of divergent outcome. *Second Language Research* 14 (4), 321–323.
White, L. and Genesee, D. (1996) How native is near-native? The issue of ultimate attainment in adult second language acquisition. *Second Language Research* 12, 233–265.

Whitman, R. and Jackson, K. (1972) The unpredictability of contrastive analysis. *Language Learning* 22, 29–41.
Williams, S. and Hammarberg, B. (1998) Language switches in L3 production: Implications for a polyglot speaking model. *Applied Linguistics* 19, 295–333.
Wölck, W. (1987/88) Types of natural bilingual behavior: A review and revision. *The Bilingual Review* 14 (3), 3–16.
Ytsma, J. (2000) Trilingual primary education in Friesland. In J. Cenoz and U. Jessner (eds) *English in Europe: The Acquisition of a Third Language* (pp. 222–235). Clevedon: Multilingual Matters.
Zangl, R. (1998) *Dynamische Muster in der sprachlichen Ontogenese. Bilingualismus, Erst- und Fremdspracherwerb*. Tübingen: Narr.
Zipf, G. (1968) *The Psycho-biology of Language. An Introduction to Dynamic Philology*. Cambridge, MA: Cambridge University Press.
Zobl, H. (1992) Grammaticality intuitions of unilingual and multilingual nonprimary language learners. In S. Gass and L. Selinker (eds) *Language Transfer in Language Learning* (pp. 176–182). Philadelphia: John Benjamins.

Index

ACT *see* Adaptive control of thought
Adaptive control of thought 44, 46-47, 56, 85, 146
Additive bilingualism 15
Ambilingual 59
Ambilingualism 92, 102
Approximative systems 107, 113-114, 160
Attractors 93, 141, 156
Autocatalytic development 107

Balanced bilingualism 117-120
Basic interpersonal communication skills 10, 16, 55
BICS *see* basic interpersonal communication skills
Bilingual competence 127
Borrowing 10, 21, 23-24, 29

CAH *see* contrastive analysis hypothesis
CALP *see* cognitive academic language proficiency
CLI *see* crosslinguistic influence
CLIN *see* crosslinguistic interaction
Codemixing 6, 21, 23-24,
Codeswitching 6, 10, 19-24, 29, 50, 59, 69, 72, 75, 92, 103, 108
Cognitive academic language proficiency 10, 16-17
Cognitive competence 28
Common underlying proficiency 17, 115
Communicative competence 32, 125-126
Competence/performance 53, 56, 58, 94, 127, 130, 145
Competence *see* language competence
Compound bilingualism 8
Contrastive analysis hypothesis 10-11, 17, 23, 25, 28, 110
Coordinate bilingualism 8
Corroboration factor 99
Critical period 7, 42, 99
Crosslinguistic influence 26, 29, 61, 98, 107, 118, 126-127
Crosslinguistic interaction 26-27, 29, 60-61, 98, 107, 118, 126-128, 130-131, 133
CS *see* codeswitching
CUP *see* common underlying proficiency

Declarative knowledge 57
Double monolingualism 57, 60, 145, 148-149
Double monolingualism hypothesis 6, 7, 18, 20, 42, 57

Effective communicative needs 135-137, 139
Emergent properties 142-143, 158
Emergent property 117, 129, 131, 157,
Emergentism 155, 157
EMM *see* enhanced multilingual monitor
Enhanced multilingual monitor 17, 129
Exponential growth 100-102

First language acquisition 20, 36, 39, 42, 55, 68, 86-87
First language system 29, 38, 43, 45, 57, 74, 86, 103, 109, 116, 118-123, 125-128, 130, 134, 140, 143
FLA *see* first language acquisition
Fossilisation 12-13, 44-46, 75, 114-115
Full attainment 127, 39

General language effort 131-137, 142
GLE *see* general language effort

Index

Grammatical competence 54

Holism 140, 144, 150-151
Homogeneous growth 89, 109, 141

Ideal types 141
Interactional competence 65
Interference 92, 94, 101, 105, 108, 110, 113, 119-120, 127, 133-135
Interlanguage 12-13, 43-45, 48-50, 65, 67, 122
Interlanguage competence 49
Invariable competence 34, 40, 109
LAE *see* language acqusition effort
Language acquisition device 34-35, 54, 57, 145-147
Language acquisition effort 122, 131-132, 134-137, 142
Language aptitude 52, 72-75, 88, 116-117
Language awareness 60, 63, 106, 108, 161
Language competence 34, 36, 38-39, 41, 44, 46, 49-50, 53-55, 57, 74-75, 88, 91, 99-100, 104-106, 114, 128, 139, 140, 145-146
Language loss 26, 76, 88, 91, 93-98, 102, 105, 135, 137, 139, 142, 159
Language maintenance 76, 92-93, 98-99, 101-102, 104, 106-107, 112-114, 129, 131, 134-135, 159, 160
Language maintenance effort 99, 101-107, 111-113, 126, 131-134, 136, 159
Language management 60, 108, 127-129, 131, 146, 149, 160-161
Language module 27, 34, 145
Language work 23, 99
Learning curve 46, 88, 100, 101, 113, 116
LME *see* language maintenance effort
LS₁ *see* first language system
LS₂ *see* second language system
LS₃ *see* third language system
LSₚ *see* primary language system
LSₛ *see* secondary language system

Measures of language loss 98
M-factor *see* multilingualism factor
MLA *see* multilanguage aptitude

(metalinguistic abilities)
MR *see* multilingualism research
Modularity of mind 32-34, 145
Monolingual competence 55, 57, 63, 115, 119, 128, 144
Monolingualism 42, 57-58, 60, 74, 103, 121
Multicompetence 49, 53, 60, 148-149 *see* multiple competence
Multilingual advantage 108
Multilingual aptitude 116-117, 126-127, 130, 138
Multilingualism factor 111, 116, 129-131, 138, 142, 160-161
Multilingual proficiency 109, 111, 127-129, 131, 133, 137, 140-141, 160
Multilingual systems 3, 19, 28, 52, 58, 64, 72-73, 86, 89, 91, 93, 111-112, 117-118, 129, 131, 140-142
Multilingualism research 52, 76, 86-87, 108
Multiple competence 160

Native-like competence 38-39, 118
Nativism 158
Negative growth 72, 91, 96, 111, 133, 135, 139

Overall communicative proficiency/efficiency 126

Paradox of transfer 14, 26-27, 61, 92, 155
Parameter switching 36
Partial achievement 43-45, 51, 114
Partial attainment 45, 98, 110, 128, 131
Partial competence 123, 160
Partial systems 45, 98, 110, 131
PCN *see* perceived communicative needs
Perceived communicative needs 91, 126, 135-137, 139,
Perceived language competence 70, 138-139
Performance *see* competence/performance
PLA *see* primary language acquisition
PLC *see* perceived language competence
Pluricultural competence 160

Positive growth 91, 133
Positive transfer 11, 14, 16, 29
Pragmatic competence 65
Primary language acquisition 39, 68-69, 116
Primary language system 122, 124, 128
Procedural knowledge 62
Proficiency 62-64, 67, 71, 74-76, 92, 96, 100-101, 107, 109, 111, 113-115, 118-120, 122-123, 126-129, 131, 133, 137, 139, 140-141, 160
Psycholinguistic turn 11, 40

Second language acquisition 6, 7, 9, 11-13, 18-21, 23-24, 26, 29-30, 33-34, 38-39, 41-43, 47, 49, 51, 53, 55-56, 58-59, 66-68, 70, 72, 86-87, 121, 123, 125, 131, 158-159
Second/secondary language system 29, 38, 43, 45, 57, 74, 86, 103, 109, 116, 118-128, 130, 134, 140, 143
Second order exponential growth 101-102, 131, 135
Semilingual 16-17
Separate underlying proficiencies 17
SLA *see* second language acquisition
Steady states 93, 97, 99, 102, 104, 106, 122-123
Strong modularity 145-146, 148
Subtractive bilingualism 15
SUP *see* separate underlying proficiencies

Theoretical monolingualism 103-104
Third language acquisition 6, 39-40, 51-52, 66-68, 73, 87, 116-117, 125, 129-131, 142, 159
Third language system 74-75, 89, 103, 116-118, 129, 143
Threshold hypothesis 10, 16, 18, 64, 111-112
TLA *see* third language acquisition
Transfer 6, 9-11, 14, 16, 19-21, 23-29, 36, 42-43, 51, 61-62, 64, 67, 75, 92, 110, 119-121, 133-135, 155
Transitional bilingualism 37, 88, 92-93, 103, 120-121, 134
Transitional competence 151

UG *see* universal grammar
Unidirectional transfer 10
Unitary competence 55, 144
Universal grammar 30, 32-44, 46-50, 55, 60, 69-70, 72, 93, 109, 143, 145, 147-148, 152

Variability (in language competence) 41, 50-52, 66, 70-72, 79, 88, 92, 96

Wave theory 71
Weak modularity 35, 145-146
Wholism 58, 148

Zipf's law 102